A SHEARWATER BOOK

One pale as yonder wan and hornèd moon,
    With lips of lurid blue,
The other glowing like the vital morn,
    When throned on ocean's wave
    It breathes over the world:
Yet both so passing strange and wonderful!

Percy Bysshe Shelley
"The Daemon of the World"

*P*assing Strange and Wonderful

ISLAND PRESS / Shearwater Books

Washington, D.C. / Covelo, California

*P*assing

*S*trange and

*W*onderful

Aesthetics, Nature, and Culture

YI-FU TUAN

A Shearwater Book
published by Island Press

Copyright © 1993 by Island Press
Jacket painting © Estate of Giorgio de
Chirico/VAGA, New York, 1992

LIBRARY OF CONGRESS
CATALOGING-IN-PUBLICATION DATA
Tuan, Yi-fu, 1930–
     Passing strange and wonderful:
aesthetics, nature, and culture /
Yi-Fu Tuan.
          p.   cm.
     Includes bibliographical
references and index.
     ISBN 1-55963-209-7
     1. Nature (Aesthetics)
2. Human ecology—Moral and
ethical aspects.   3. Human
geography—Moral and ethical
aspects.   I. Title.
BH301.N3T83   1993
111'.85—dc20   92-37105
                              CIP

Manufactured in the
United States of America

10 9 8 7 6 5 4 3 2 1

To my brother Tai-Fu

CONTENTS

Prologue: Light and Shadows / 1

PART I: AESTHETICS, NATURE, AND CULTURE

Chapter 1. The Aesthetic in Life and Culture / 5

2. The Development of the Aesthetic Impulse / 20

PART II: SENSORY DELIGHTS

3. Pleasures of the Proximate Senses / 35

4. Voices, Sounds, and Heavenly Music / 70

5. Visual Delight and Splendor / 96

PART III: A CULTURAL-AESTHETIC SPECTRUM

6. Australian Aborigines, the Chinese, and
Medieval Europeans / *121*

7. American Place and Scene / *143*

PART IV: SYNTHESIS OF SPACE AND THE STATE

8. Synesthesia, Metaphor, and Symbolic
Space / *165*

9. Ritual and the Aesthetic-Moral State / *182*

The Chinese Aesthetic-Moral State: T'ang Empire / *186*

Renaissance Venice as an Aesthetic-Moral State / *193*

France: The Sun King's Aesthetic-Moral State / *196*

America the Beautiful and the Moral / *199*

PART V: AESTHETICS AND MORALITY

10. Good and Beautiful / *213*

Epilogue: Shadows and Light / *227*

Notes / *245*

Acknowledgments / *271*

Index / *273*

Passing Strange and Wonderful

Light and Shadows

*T*his book is about the great importance of the aesthetic in our lives, and about the different forms it takes in a spectrum of cultures. Many people may feel that although beauty does matter, it is an "extra," something we like to have in our surroundings when more basic needs are met. Yet the pervasive role of the aesthetic is suggested by its root meaning of "feeling"—not just any kind of feeling, but "shaped" feeling and sensitive perception. And it is suggested even more by its opposite, anesthetic, "lack of feeling"—the condition of living death. The more attuned we are to the beauties of the world, the more we come to life and take joy in it.

It is possible, then, to read the book as an orderly presentation of aesthetic experiences—a guide to the splendor of the earth and of the human creations on it. But I hope the book offers more. I see it as opening up a new view of culture and the aesthetic, such that the aesthetic is taken to be not merely a dimension or aspect of culture, but its emotional-aspirational core, both its drive and its goal.

The book moves from the simple to the complex: from a consideration of the different human senses separately to the commingling of sensory responses (synesthesia), from the mental blending of images and concepts as in metaphors and symbols to the creation of symbolic spaces, and finally to aesthetic-political states. This approach emphasizes the developmental aspects of culture and aesthetics. Notwithstanding the skepticism of many scholars regarding the idea of progress, it is sometimes necessary to speak of cultures in such developmental terms. Changes in the direction of expansion or elaboration, new ways of perceiving and making rather than failures of imagination and the decay of society and culture, human ingenuity rather than human folly, are the core of my story.

Most recent works on environment and society tend to stress the dark side of things: pollution, exploitation, greed, and the like. By contrast, the outlook presented here is predominantly sunny. Yet an undertone of unease threads the book, if only because every road followed implies other roads not taken, every new creation implies a prior stage of destruction, and every new perception dims, if it does not wipe out altogether, the old, which has its own—perhaps irreplaceable—value.

I

Aesthetics, Nature, and Culture

The Aesthetic in Life and Culture

I wake up, open the curtains, and am confronted by a landscape lit by the early-morning sun. Soon the sound of birds, the cool touch and fragrance of morning air envelop me. The bright and dewy presence of the world makes it seem newly created, and makes me feel good to be alive. That phase quickly passes. Practical demands take over. The first port of call is the bathroom—the locale for transforming animal into person, body into social being, nature into art. Arranging that wilderness of hair into some semblance of order is the original meaning of cosmos or cosmetics. Next stop is the kitchen. I wipe the kitchen counter clean, but note that some coffee grounds have wedged themselves into cracks and would be difficult to remove. On the blue-rimmed china plate is an egg sunny-side up, a golden orb in the field of white, which I break reluctantly with a fork and then quickly eat to remove the evidence of an offense against perfection.

At the office, too, my mind takes in, at some level, the physical environment: the books on the shelves, their brightly colored jackets in the light at different times of the day; the ill-fitting elevator door; the lukewarm water from the cooler. Much of my time and the time of my coworkers is spent in performing—and also judging— social rituals. The penalty for poor performance is not a dock in pay or social ostracism, but rather the nagging feeling of having missed a cue—of a certain awkwardness in one's bearing and conduct toward others that reflects adversely on one's image of oneself. We are all on stage, for at some level of consciousness I also judge other people's facial expressions, gestures, and behavior.

Four colleagues fail to turn up at a scheduled meeting, leaving gaps around the oval table that make communication a little more difficult. The gaps seem to me like missing teeth in an open mouth. One man, I observe, makes a point succinctly and eloquently; another offers a rambling, interminable commentary. Exchanges sometimes soar; more often they crawl. The sparks of wit relieve the drowsy stolidity pervading the room.

Lunch is arranged in honor of a colleague who is about to take a two-year leave. After it, we rise from the table, shake his hand, wish him well, and tease him about having to surrender his coveted parking space. We give him a final pat on the back. In a film, the scene fades at this point. In real life, however, we make our way to the coat-rack, pick up our many paraphernalia for protecting ourselves against the winter cold, and straggle to the elevator, where we meet again, this time unplanned, and wait awkwardly in each other's presence without a prepared script, until one person has the wit to say, "Such ragged endings are the stuff of real life." By making this remark he has saved the appearance; he has transformed life into art.

As the vignette suggests, culture has a variety of meanings. Culture is a physical process that changes nature. I rearrange my unkempt hair into a semblance of order; I break an egg to cook sunny-side up. On another scale, pioneer settlers clear forests to create farms.

Culture is perception. I see the beauty of dawn, savor the fragrance of morning air, and note with annoyance the ill-fitting door of the office elevator. On another level of sophistication, I recognize (with the help of Impressionist painters) the beauty of dappled sunlight on the facade of a cathedral.

Culture is speech. Humans use language not so much to convey factual information as to construct worlds, illuminating certain facets of reality while throwing others into the shade, and calling up images that demand appraisal and sometimes action. A story well told—even if it relates a scientific experiment or a business venture—is successful rhetoric, an artful arrangement and articulation of words that strike their listeners as convincing. Speech, moreover, affects its material environment: a dull, interminable speech can make even the room's furniture seem drab, whereas a witty remark lightens the air, imparting an extra sparkle to the common water glass.

Culture is performance—facial expression, gesture, and social ballet. Image and show permeate our society as they have all others, in different ways depending on time and place. Moralists see this ubiquitous urge to present the self as a sort of disease, a fall from a state of grace, a loss of some mythic golden age when people were, in some unspecified sense, genuine, when they did not put on an act. And yet it was not a twentieth-century public relations expert who produced the words "All the world's a stage." The theatrical model of human reality is deeply ingrained in Western thought. And if this model is not as strongly and persistently stressed in other cultures and civilizations, it may be simply because it has been found too obvious to need to be put forward overtly.

The Aesthetic, Consciousness, and Emotion

The aesthetic impulse, understood as the "senses come to life," directs attention to its roots in nature. But though rooted in nature (biology), it is directed and colored by culture. Indeed, the ability to appreciate beauty is commonly understood as a specialized cultural competence, which varies from individual to individual and from group to group. How do the meanings of nature, culture, and the aesthetic differ, and, in particular, what is there about the aesthetic that makes it deeply a part of nature and culture and yet also endowed with traits uniquely its own?

Most societies distinguish between nature and culture and consider that distinction important. In general, wherever the distinction is recognized, the biological, the raw and the instinctive, the unconscious and the primordial are attributed to nature; and form and order, consciousness and deliberation, the developed and achieved ideal are attributed to culture. Although the distinction is clear at a conceptual level, nature and culture so interpenetrate that it is often hard to say of an object or behavior whether it is more the one than the other. Culture, through habitude, easily becomes second nature—like the graceful gesture that feels natural, always there, rather than chosen or constructed. Stable customs and institutions, too, often fade into the background of consciousness, like the unvarying cycles of nature.

The level of consciousness, then, is an indicator of that which distinguishes not only between nature and culture but also between culture and the aesthetic. Cultural activity is, in varying degrees, conscious activity. At the beginning of any new project, consciousness is at its peak: people have to envisage actively what is functional, right, and appropriate, as against what is dysfunctional, wrong, unnecessary, and ugly. When there are frequent pauses for appraisal and appreciation, cultural activity is also aesthetic activity. Most practical cultural activities sooner or later become routine, however, and then people are barely aware of what they do. If the task is complicated, pauses may be necessary to consider the next step; but such pauses are part of a goal-oriented action, merely to permit practical thinking. But if people use the pauses to savor what they do and envisage the "perfection" that is yet to come, they are being actively cultural—cultural in the aesthetic mode.

For the aesthetic mode to occur requires a certain distancing from the flow of life, the embeddedness of nature, the routines of culture. It is a mood, a feeling, an emotion. But these cannot take extreme forms: violent emotion (such as rage or lust) destroys the requisite psychological distancing; drowsiness plunges one back into the enveloping bosom—the embeddedness—of nature. What are these extremes, why do we reject them, and which emotions lie within the range of true aesthetic experience?

SLEEP, DROWSY INDOLENCE, DREAM

Consciousness is opposed to oblivion, as aesthesia is opposed to anesthesia. In sound sleep one is oblivious not only of the world but of sleep. Sleep itself cannot be savored and thus is not an aesthetic experience. When a young, healthy person stretches her arms and smiles into the morning sun, claiming to have enjoyed a good night's sleep, she is savoring the effects of sleep—the sense of well-being and restedness—and not sleep itself. The moment before one enters sleep's oblivion can also be enjoyed. Montaigne found the silk-smooth slide into unconsciousness so rewarding that he asked his valet to wake him early so that, realizing that it was still early, he could repeat the experience.[1]

But the voluptuous sense of drifting into sleep cannot properly be called an aesthetic feeling, because the enjoyment experienced is not of something external to the self: it is not a state of being that can be learned and perfected under the aegis of art and culture. The same exclusion applies to the state celebrated in Keats's remarkable "Ode on Indolence":

> *Ripe was the drowsy hour;*
> *The blissful cloud of summer-indolence*
> *Benumb'd my eyes; my pulse grew less and less;*
> *Pain had no sting, and pleasure's wreath no flower:*
> *O, why did ye not melt, and leave my sense*
> *Unhaunted quite of all but—nothingness?*

In a state of drowsy indolence, one is still alert enough to enjoy all kinds of soothing, happy sensations, to feel the warmth of the sun and the fragrance of the flowers. Only the visual world is curtailed.

"Steep'd in honied indolence" (as Keats puts it elsewhere), a man is not sufficiently removed from his immediate environs to appreciate a view "out there." Perceptions are blurred, have become "soft." Not only has pain "no sting," but pleasure's wreath has "no flower"—no image of an external reality. Out there is "nothingness."[2]

The American writer Paul Goodman offers another picture of indolence. Confronted by the foreignness of Italy, he feels slightly uncomfortable with the unfamiliar place and language. "But when the whirlpools of the wake turn after my dripping oars, and I lie down in the belly of my rowboat in the sun, I am as much at home as on Lake Seneca, a citizen of nowhere but an animal of the world. I am lucky to be able to come so far and end up in the same place."[3] There is no scenery to be appreciated from the belly of a rowboat. There is only that sense of animal well-being, which is placeless, below culture and aesthetics.

Certainly, sound sleep is oblivion. But is a pleasant dream during sleep an aesthetic experience? The dreamer's "landscape" is often a mood, induced eerily by a particular feature (house, tree stump, dead bird) rather than by a topography. Even when the dreamscape seems to have a distinctive topographic character, the dreamer lacks the ability mentally to remove the self. Awake, one can easily achieve a sufficient distance to say, "Life is like a dream," but in a dream one is not free to say, "It is like being awake." In a dream, one lacks the power to think even at an elementary level—to say, for instance, that although one is here, one could be elsewhere, and that although the sun is shining now, it could rain next week. In short, dream is immersion: the dreamer is a captive of the milieu and time in which she finds herself. This bondage to the immediate environment is especially unyielding in a nightmare. Its singular horror consists in the tight grip of events, the subjection to a world of palpitating evil that does not allow for any redeeming, normal moment, such as noting from the corner of the eye a man sweeping up leaves.

RAGE AND HATE

At the other extreme, powerful emotions obliterate the distance between self and other. They overflow boundaries and destroy reality's freestanding existence. Anger is volcanic or explosive; rage is a torrent, directed perhaps at an individual but destructive to all who happen to stand in its way; sexual lust is all-consuming, violent, and indiscriminate. Strong human emotions are readily likened to forces of nature. The heroes of Homeric Greece, for example, were likened in epics to nature on the rampage—roaring like thunder or a lion and sweeping all before them as in a torrential flood. Heroes were proudly mad. "Fighting madness," writes classicist Jasper Griffin, "is no mere figure of speech. Impelled by Zeus in an outburst of irresistible fury, Hector foams at the mouth, his eyes flash, and his helmet shakes terrifyingly as he fights. The word used of the raging hero is often the word used of the madness of a mad dog." In the Homeric world, cannibalism and the eating of raw animal flesh are permissible reversions to primitive nature. Achilles says to the dying Hector, "Would that my passionate heart would incite me to chop up your flesh and eat it raw." Hecuba in her turn longs to devour Achilles' liver in revenge for Hector's death. These practices have parallels in Germanic culture and were, according to Griffin, "genuinely ancient Indo-European ideas of the ways terrible heroes behaved."[4]

Outbursts of anger remained socially acceptable in Western culture until modern times. They were not considered a serious disruption of life even in good society. One's face might be distorted in wrath, but call it righteous wrath and it would be deemed justified, indeed commendable. During the eighteenth century, displays of anger became somewhat less acceptable in the home; the word *tantrum* came into use, and the word *temper* took on a negative meaning. In the Victorian period the repression of anger began in earnest: the family, in particular, was regarded as a haven of peace, and anger among family members was to be avoided at all cost.[5]

Rage and other forms of violence are beyond the pale of the aesthetic. A cultural history of the West can be written as the story of the refinement of manners, the progressive control of strong emotions, and the eschewal of displays of visceral or animal violence. Delay is a key to the refinement of manners—the cooked rather than the raw, the exploration of a roast with a knife and fork rather than the plunge into it with bare teeth and hands. Pause is a key to the control of emotion: the tide of feeling is momentarily stayed so that it can be channeled into more artful, if just as deadly, ways—the dismissive look rather than the shout of rage, the cold steely eye rather than the foaming mouth.

Hate differs from rage and lust in that it is directed to a specific object in the external world. Hate is not just a swooning or explosive emotion, and perhaps for that reason it may be characterized as "cold." Simone Weil provides an example: Suppose I hate a man; as he approaches, something hateful—a loathsome quality—approaches me. "If he is a blond, it is a hateful blondness, and if he has brown hair, it is a hateful brownness."[6] A color can acquire a hideous vividness through the lens of hate. Hate attends to detail in a way disturbingly like that of love. John Updike shows how hate can call into play the discrimination and refinement of an aesthete.

> And he would hate him—hate his appearance, his form, his manner, his pretensions—with an avidity of detail he had never known in love. The tiny details of his roommate's physical existence—the wrinkles flickering beside his mouth, the slightly withered look about his hands, the complacently polished creases of his leather shoes—seemed poisonous food Orson could not stop eating. His eczema worsened alarmingly.[7]

Hate, nevertheless, is not an aesthetic emotion, and the reason becomes clear when we contrast it with love.

SEXUAL DESIRE, LOVE, EROTICISM

Susan Sontag characterizes sexual desire as "one of the demonic forces in human consciousness—pushing us at intervals close to taboo and dangerous desires, which range from the impulse to commit sudden arbitrary violence upon another person to the voluptuous yearning for the extinction of one's consciousness, for death itself."[8] At a certain level of intensity, sexual passion surges beyond the aesthetic. Below that level, when passion tempers to desire or longing, eroticism energizes the feeling for beauty, giving the loved object a glow—a palpable warmth—that is its life-stirring power.

Eroticism, moreover, vastly expands the range of objects that can exert a riveting appeal. Human beauty is inescapably tinged by erot-icism, if it is genuinely felt and not a response dictated by conven-tion. A baby's shape, its milky fragrance, the softness of its skin, the gurgling music of its attempted speech, all make it an irresistible target for adult admiration. Almost all parts of the human body can have enormous aesthetic-sexual appeal. Women's eyes and hair have evoked paeans from Western poets for millennia. But one can also be drawn to one's "mistress's eyebrows," to a semitranslucent earlobe, slightly tipped nose, sculptural lips, strong masculine fingers, and even the feet. A lover sees details invisible or totally uninteresting to the profane eye. A character in John Osborne's play *A Patriot for Me* declares: "I tell you this: you'll never know that body like I know it. The lines beneath his eyes. Do you know how many there are, do you know one has less than the other? And the scar behind his ear, and the hairs in his nostrils, which has the most, what colour they are in what light?"[9]

Although hate can be just as discriminating and attentive to de-tail, the psychological distance between hate and love is immense: it is the difference between enthrallment and enchantment, between being enslaved—a life-diminishing force—by "the wrinkles flick-ering beside his mouth"—and being enchanted—a life-enhancing force—by "the lines beneath his eyes." It is the difference between the nonaesthetic and the aesthetic, between capture that is impris-onment and capture that, paradoxically, enlarges and liberates.

Just as the aesthetic appeal of the human person is inevitably touched by the glow of eroticism, so, as an effect of the metaphoric operations of the human mind, is the aesthetic appeal of many to-pographical features. Think of how often we anthropomorphize such features: foothills, a headland, the mouth of a river, the eye of a storm, the face of a cliff, the brow of a ridge, the shoulder of a valley, an arm of the sea. As we look at the "shoulder" of a valley, our eyes following its strong bold curvatures, there can be no denying an erotic tinge in our appreciation.

AESTHETICS OF THE ABSTRACT

Overwhelming emotion does not allow the distancing necessary to aesthetic experience, nor does a condition of repose in which consciousness, lulled by gentle sensations, hardly reaches out to the world. If the distance is too great, the pause too long, the experience will tend to be intellectual rather than aesthetic for lack of intimate, warmth-generating contact with the sensory realities. But cannot the intellectual also be aesthetic? Is it true that intellectual understanding is devoid of warmth? Does it always lack those changes in bodily state (the sparkling eye and the tingling skin) that are the accompaniments of emotion? To Roland Barthes, for one, "abstraction is in no way contrary to sensuality." He finds it natural to associate intellectual activity with delight: "the panorama, for example—what one sees from the Eiffel Tower—is an object at once intellective and rapturous: it liberates the body even as it gives the illusion of 'comprehending' the field of vision."[10]

From the top of the Eiffel Tower, people look like ants—moving dots, abstractions. Reality, from that height, is map rather than scene or landscape. One is removed both physically and psychologically from the concerns and turmoils of life. The loss of warmth and heat in direct human involvement is compensated by contemplative gains with their own emotional temperature and quale. One gain is a sense of the sublime, as a person moves from the familiar multisensory spaces to what Marshall McLuhan calls the "vast, swallowing distances of visual space."[11] Here is how Edgar describes to blind Gloucester the view from the edge of a cliff (*King Lear*, Act IV, scene 6):

How fearful
And dizzy 'tis to cast one's eyes so low!
The crows and choughs that wing the midway air
Show scarce so gross as beetles.

Rather than dread, the distant view can give one a sense of power over the earth, with its components spread below for the enjoyment of the imperial eye. Mountains have always been there to provide the distant view, but in premodern Europe few people climbed them for such a purpose. Even in Shakespeare's time the space that yawned below the edge of the cliff was more likely to generate fear than a sense of command. Eventually, mountain climbing found favor among the intrepid, as also the panoramic spreads it offered. Flying gave the same sort of satisfaction, whether in a hot-air balloon (invented in the late eighteenth century) or in a sleek aircraft of our own era. To those disinclined to climb or fly, the skill of cartographers has made godlike views of the earth readily available. Terrestrial mosaics revealed first by topographical maps, then by aerial photographs, and now by satellite imagery are admired not only for their wealth of information but also for the strange beauty of their abstract patterns: what is familiar on the ground and barely worth a second look can become unfamiliar and beautiful when seen from above.

Finally, the distant view can be fond or tender rather than cool. Emotion has not disappeared; rather, it has taken on a different quality. A Chinese poem of the fourth century B.C., inspired by shamanism, presents the point of view of a man in flight from the earth, striving toward the splendor of the heavens:

> *But when I had ascended the splendor of the heavens,*
> *I suddenly caught a glimpse below of my old home.*
> *The groom's heart was heavy, and the horses for longing*
> *Arched their heads back and refused to go on.*[12]

In Shakespeare's *Richard II*, John of Gaunt, contemplating England as though from a towering height, describes it affectionately as "this little world, this precious stone set in the silver sea" (Act II, scene 1). Astronauts develop a tenderness for the Earth, with its beautiful and fragile sheath of life, threatened by pollution—smoke from cities and forest fires, the smears along the coast where dirty rivers enter the ocean—that is clearly visible from their spacecraft. The Russian astronaut Georgei Grechko, who has flown for a total of 3,200 hours, says: "I noticed in space a certain shift in perception—you begin to value things that are taken for granted on earth, like fresh air, a mountain stream, sunrise in a forest, a shivering of leaves on the trees—all these things we do not pay much attention to down here. In the space station, one picture we chose to put on our wall was of an ordinary maple leaf."[13]

From a high mountaintop, an airplane, or a spacecraft, one's perception of the earth is abstract in the sense that it is limited to what the eyes see rather than to what all the senses apprehend. Beyond seeing with the eye, a further step in the direction of the abstract is seeing with "the mind's eye." The sexual impulse that lends warmth to all sensory-aesthetic experience would seem incapable of penetrating the abstract mathematical realm. Yet Alan Turing, perhaps in a teasing mood, claims to derive sexual pleasure from mathematics.[14] Many physicists and mathematicians have maintained that a sense of beauty is the ultimate guide to the significance and truth of their work. Laypersons are likely to think of this aesthetic appeal as almost entirely cerebral and hence best described as "distant" and "cool." But scientists themselves use stronger words. S. Chandrasekhar speaks of his moments of mathematical success as "shattering," "profound," "shuddering." The mathematician G. N. Watson uses the vivid French expression *chair de poule* for a sense of the uncanny. He responds at times in visceral awe, but also with considered admiration. Srinivasa Ramanujan says that an equation can give him a thrill comparable to the feeling he experiences when he contemplates the beauty of certain sculptures by Michelangelo.[15]

Bertrand Russell perhaps puts it best. For him, emotional intensity is not the issue. Mathematics, he says, is "capable of an artistic excellence as great as that of any music," not because the pleasure it gives is comparable in intensity, "but because it gives in absolute perfection that combination, characteristic of great art, of godlike freedom, with a sense of inevitable destiny; because, in fact, it constructs an ideal world where everything is perfect and yet true."[16]

Culture as
Material Transformation

Cultural history, I have noted, can be written as a story of the refinement of manners: raw passions yield to subtler sentiments and feelings so that, for some people, even mathematical equations can provide an aesthetic thrill. But cultural history can also be written from another point of view—as the material transformation of nature. Anthropologists at one time defined the human species as *Homo faber*. Our distinguishing trait is that we make things. Although that label has since been discarded as unscientific, the popular view still takes our uniqueness to lie in the "secondary world" we have erected on the primary one of nature, and at the expense of nature.

Much that we make is functional and economic—necessary to the maintenance of a livelihood. And economic activity is generally regarded as nonaesthetic or even unaesthetic. Yet when raw material or amorphous matter is given shape, when clay is made into pot, the enterprise can only be described as artisanal-artistic. It is hard to see how the aesthetic impulse can be excluded from a psychologically true account of the making of even the most mundane things— even, say, a drainpipe. The end product *pleases* if it performs as planned; and it is a very short step from "pleases" to "pleasing," an adjective that has an inherent aesthetic meaning.

On one scale, artisans make tools and implements, household utensils and furniture, to support a way of life. On a much larger scale, agriculturalists clear bush or forest to produce a world of their own. The sustained effort, if we can observe it, would seem anything but aesthetic. Wherever it occurs, in a tropical forest or in nineteenth-century America or Australia, it is arduous beyond imagining. "Every beginning," Goethe tells us, "is difficult." An Australian historian applies this dictum to his own country. "In Australia," he notes, "every beginning has not only been difficult, but scarred with human agony and squalor. Mountains of suffering, material hardship, and a cultural semi-desert were the prices paid to begin the pastoral industry. The economic activities begun after 1850 continued in this tradition. Agriculture in its birth pangs . . . spawned still more squalor, more wretchedness and more human suffering."[17]

In every pioneering venture, survival is at first the predominant concern. But both on the large scale and in the minute details of daily life, it is hard to imagine that people vigorous enough to take on the new have no moments of hope in their efforts to create order, or moments of delight in their surroundings. Pioneer farmers must in some sense envisage the good life—form pictures, however hazy, of the kind of farm they will eventually have and of the satisfactions to come. That act of the imagination is aesthetic. Even in the harsh environment of the frontier, people struggle not merely to survive another day, but also to create and maintain a world of order and meaning, be it a swept corner or a clean shirt, a framed picture, or other objects transported from the old home to the new at considerable cost. These mementos of a familiar past have little or no practical value, but they provide real emotional and psychological as well as physical comfort in the raw environment. People need some token of craftsmanship and beauty in almost all circumstances of life.

In contrast to the rawness of the pioneer farm, a mature agricultural landscape can have undeniable charm and beauty: a Wisconsin farm with tall corn, stately silos, and handsome barn; the glorious vineyards of the Loire Valley in France; the rice fields of south China, rising tier after tier in sculptural splendor. These created environments attest to the aesthetic flair—the unself-conscious exercise of skill—in works that are primarily economic. And a great city is a still more conscious exercise in striving for aesthetic excellence, in its temples, palaces, sculptures and paintings, public squares and tree-lined avenues, museums and department stores, fountains and gardens. It can be thought of as a complex of nested containers, of jewel boxes within jewel boxes, down to the drawer of a dressing table in a modest house, inside which is an enameled sewing box containing some pretty opalescent buttons.

Moral Dilemmas

When artwork attains the scale of monuments in a city and of the city itself, we are forced to face the fact that the aesthetic intersects with the moral and can enter into violent conflict with it. A desire for the pleasing object and experience, for a physical setting that both stimulates and soothes the senses, is innocent enough; it is, in any case, human, and culture is inconceivable without its motivating power. When great desire combines with great talent, the result is the dedicated artist. Dedication calls for personal sacrifice, but if the sacrifice is willingly offered, no conflict with morality occurs. The moral question comes to the fore when a division of labor exists between designers and executors. The potentate who commands the work does no work at all; yet we say, "Kublai Khan built Cambaluc." The artist-architect seldom toils except over the drawing board. The numerous laborers who execute the design sweat and toil, but they get little or no credit, and the product is seldom theirs to enjoy. Moralists, confronted by such creations, tend to see embodiments of social injustice rather than beauty: the moral reading overshadows the aesthetic.

Aestheticism has also been criticized for two reasons that appear to be contradictory. On the one hand, its enjoyment of transient and disconnected moments is considered irresponsible: the aesthete takes delight, like a child, in a balloon, the striped awning of a café, a towering construction crane, the fragrance of burning leaves, a military parade, or a sunset without any attempt to see how they might constitute a larger whole and without examining the consequences they may have on human life, society, and nature. On the other hand, aestheticism, backed by sufficient power, is tempted to assume the prerogative of God in total creation. The cosmetic effort that begins modestly as a rearrangement of one's hair or as the paving of a footpath in one's garden may end, in the megalomania of a dictator-aesthete, in the desire for a total transformation of society and nature. Clearly, the pursuit and enjoyment of beauty are full of paradoxes. Beauty can both enhance the moral by infusing it with sensual appeal and mask it with glitter. The aesthetic attitude, in contrast to the practical one, presumes receptivity and a disposition toward inaction, yet it can also be a driving passion for control.

Extreme aestheticism is a sophisticated attitude—a cultivated posture toward life and world. No one is born an aesthete. Yet the aesthetic impulse is innate. We all have it in some degree. Every society nurtures this impulse in its young, encouraging it to grow in scope, power, and subtlety. The specific artifacts and arts thus produced may vary greatly from culture to culture. Less varying are the child's stages of development in aesthetic competence and appreciation, which are also those of mental and physical maturation. It is to these developmental stages that we turn in the next chapter.

CHAPTER 2

The Development of the Aesthetic Impulse

*I*n all societies that explicitly distinguish between nature and culture, children are considered to begin as creatures of nature and to move progressively toward culture as they are taught to accept the values and achieved forms of society.

Certainly, very young children seem to be ruled by animal instinct. Asleep, her bodily needs satisfied, an infant looks angelic; awake, she can suddenly become a noisy bundle of rage, fists clenched, legs kicking, howling furiously, her entire body turning lobster-pink. Young children care little about their own appearance, and their tantrums are as unpredictable, sudden, and violent as summer squalls. Indeed, from one perspective, young children experience their effectiveness in the world through creating chaos out of order. If they appear tidily dressed, well mannered, and appreciative of adults' orderly world, it is because they have been so taught, patiently and often against a background threat of force.

Another view, urged strongly in modern times and probably colored by adult retrospection, is that young children live in a world of wonder—that, far from being insensitive to beauty, they have a privileged access to it.

To the unprejudiced eye, it is clear that even very young children are endowed with an aesthetic impulse, by a sense of satisfaction with and delight in certain configurations of reality. Newborns prefer certain sounds and rhythms over others: they tend to relax and stop crying when they hear recordings of heartbeats or of the womb sounds of blood pulsing through various maternal blood vessels. They also show visual preferences. In one experiment, a six-week-old infant was linked up with a sucking mechanism, which in turn was connected to a projector. By sucking with varying speed the infant could cause a picture on the screen to become either blurred or clarified. The researcher, J. S. Bruner, found that "a good visual stimulus, concentrically organized and sharply contoured, will have the effect of inhibiting sucking altogether," and concluded from this result that "the epistemic needs of the newborn organism are not completely swamped by the need for food and comfort."[1] What pleases the newborn is a combination of roundness of form and clarity. Not only the epistemic but also the aesthetic needs of the infant demand satisfaction; the two are closely intertwined. More recently, researchers have found that infants six to eight months old prefer "pretty faces"; they look longer at attractive faces (as judged separately by adults) than at unattractive ones, regardless of the physical appearance of their mother. Experiments of this kind suggest that the concept of beauty is not determined solely by culture.[2]

If we think of children as egocentric, emotional, and constantly in motion, we will tend to give minimum credit to their capacity for aesthetic experience. We notice, for instance, that young children rarely appreciate landscapes. When adults get out of the car to admire the panorama, youngsters either rush about in enjoyment of their freedom or feel bored. However, if we view young children not as hyperactive and egotistically involved in their own urgent needs and projects, but simply as unself-conscious, intensely aware, and unburdened by the numerous goal-directed activities of adults, we tend to give much credit to their capacity for aesthetic experience.

Timelessness
and Sensory Delight

Many young children appear able to dwell in an expansive timeless present not available to adults, nagged by practical tasks. How does one characterize this state? Is it a luxurious feeling of indolence, edged distantly by the threat of boredom? Is it a sensual bath in the rich, unorchestrated stimuli of the environment—the sound of bees, the smell of hay and fertile earth, the sensations of an ant crawling up one's leg and of the sun gently baking one's chest? Such a concourse of experiences is not exactly aesthetic: the distance that allows reflective appreciation is lacking; the sensory and the sensual overwhelm the intellectual.

The transition from sensation to aesthetic appreciation is a subtly graduated change that depends, along with other factors, on the degree to which an external reality exists for the self. At one extreme is a dreamlike immersion in sensations; at the other is the active (critical) enjoyment of scenery. Young children are particularly prone to dwell somewhere between these two. One environment that promotes this intermediate mode of being is the nook. A nook can be the tunnel behind the living room sofa, the space under the grand piano, but most satisfyingly it is an enclosure outdoors—a natural cave, a hollow in the overgrown bush, a tree house. Children all over the world have shown a fondness for the nook. Its special merit lies in offering at the same time a cozy space suited to the size of the child and a beckoning openness beyond. The womblike hollow on the one side and open space on the other capture for the child, in a manageable and comprehensible way, the basic polarities of life: darkness and light, safety and adventure, indolence and excitation, multisensory ease and visual alertness.[3] For the child nestled there, time comes almost to a stop: past and future cease to exist, displaced by a transcendent present. Robert Grudin recalls such a childhood idyll—a calm happiness, nurtured by an enveloping profusion of pleasant stimuli: "I taught my younger brothers to climb . . . a tree behind the house in Red Bank where my parents still live. We would scramble up the old trees, sit on the vine-upholstered benches that were their tops, munch on berries and otherwise do nothing. Here we developed . . . a love of total idleness, a taste for enjoying, silent and motionless, the isolated and retarded time which rose all around us like a new element in the warm stillness."[4]

Capacity for Wonder

By virtue of their immense natural vitality, children are also more likely than adults to possess an acute sense of wonder, an intense openness to the world. This capacity presupposes a distance between the self and the nonself—a recognition of the strange and marvelous other. Children do not yet feel quite at home in the world. They have not yet had time to establish the routines or embrace the interpretative schemata that can make the world seem predictable, familiar, even gray. Events and objects stand as it were in the bright sun, alone, beyond the shadows of competing events and objects. According to Richard Coe, the "challenge to the intuition of the child . . . is precisely that which *cannot* be assimilated to the human: centipede or stamen or sea anemone; pebble or porcelain button or crystal peardrop fragment from a vanished chandelier. . . . In the subconscious aesthetic of childhood . . . the beautiful imposes itself with the irrational violence of arbitrary fact."[5]

Remembrance of Things Past

Color and sparkle seem, above all other qualities, to have the power to waft the young into their magical world: "The forever inaccessible, strange beauty at the heart of the glass marble; the fragile, iridescent loveliness of the soap bubble; the wonder of the kaleidoscope, transforming arbitrary slivers of colored glass into patterns as meaningful as the stained-glass windows of the Sainte-Chapelle; the enchantment of transfers—greyish, opaque chrysalises, out of which emerge as if by magic the multihued contours of the final images."[6] Vladimir Nabokov recalls that as a toddler he once played in his crib with "a delightfully solid, garnet-dark crystal egg left over from some unremembered Easter. . . . I used to chew a corner of the bedsheet until it was thoroughly soaked and then wrap the egg in it tightly, so as to admire and re-lick the warm, ruddy glitter of the snugly enveloped facets that came seeping through with a miraculous completeness of glow and color." At age four, traveling abroad with his parents on the Mediterranean Train de Luxe, he remembers "kneeling on my (flattish) pillow at the window of a sleeping car . . . and seeing with an inexplicable pang, a handful of fabulous lights that beckoned to me from a distant hillside, and then slipped into a pocket of black velvet."[7]

Glitter is magical. But other qualities can seem magical too. For the young Pierre Teilhard de Chardin, dull heavy iron captured the quiddity of substance, completion, and plenitude.[8] The young C. S. Lewis found magic in a miniature landscape. "Once in those very early days my brother brought into the nursery the lid of a biscuit tin which he had covered with moss and garnished with twigs and flowers so as to make it a toy garden or a toy forest. That was the first beauty I ever knew. . . . It made me aware of nature—not, indeed, as a storehouse of forms and colours but as something cool, dewy, fresh, exuberant."[9]

Lewis seems to have accepted the idea that the aesthetic applies primarily, if not solely, to form and color. He recalls that he and his brother, at ages six and nine respectively, liked to draw.

From a very early age I could draw movement—figures that looked as if they were really running or fighting—and the perspective is good. But nowhere, either in my brother's work or my own, is there a single line drawn in obedience to an idea, however crude, of beauty. There is action, comedy, invention; but there is not even the germ of a feeling for design, and there is a shocking ignorance of natural form. Trees appear as balls of cotton wool stuck on posts, and there is nothing to show that either of us knew the shape of any leaf in the garden where we played almost daily.[10]

The issue of whether children are aesthetically inclined is confused by the adult tendency to judge their preferences and abilities in accordance with adult standards. Thus Lewis argues for his childhood self's almost total lack of aesthetic sense; yet what is drawing but a desire to create form? And if the child Lewis can draw movement—a difficult accomplishment that does not appear overnight—why does the adult Lewis say so confidently that in his youthful drawings "there is not even the germ of a feeling for design"? A further source of difficulty in settling the issue of children's aesthetic impulse is the difference between the recognition of beauty and the ability to represent it. A child recognizes the magical beauty of the lights from the windows of a passing train at night, but how can he hope to represent it in words or pictorially? Even a very young child may be able to appreciate the aesthetic accomplishments of an older person; she can see that a picture or a performance is better than what she can produce. As a consequence, she may feel humiliated. John Holt describes a little girl not yet a year old who "had been given a little plastic whistle, which she loved to toot. It was her favorite toy. One day one of her parents picked up the whistle . . . and began to play a little tune on it. They both amused themselves with it for a minute or two, then gave the whistle back to the baby. To their great surprise, she pushed it angrily aside."[11]

Growth in Competence: Music and Visual Art

Children develop aesthetic competence in remarkably similar stages, whether the medium is music or pictorial art.[12] With regard to music, children up to age three seem to appreciate primarily the tone itself rather than a rhythm or melody, and they show a desire to reproduce it—to master it. Very young children, for instance, like to pluck the string of a guitar, and listen to the sound from the start to its dying resonance. By age three or four, they like to experiment with acoustical range, showing "a keen interest in very soft and very loud sounds; a loud bang on a drum accompanied by expressions of sheer delight or even fear, or a fascination with the very gentle sounds of a shaker or Indian cymbal."[13] They may simply play up and down an instrument's scale and produce musical fragments, repeating them over and over. But they may also experiment, deliberately change levels of speed and loudness, and use larger or smaller intervals for expressive purposes. For example, one four-year-old girl attempts to produce a song in response to the idea of the sun shining. She is not trying to "illustrate" sunshine or a heavenly body; rather, she seems to want to express a feeling of inner glow—the musical equivalent of a "sunny disposition." The song thus produced makes little use of any other songs she may know. It is original, personal, and idiosyncratic.[14]

From about age five to age eight, children's musical interest centers on imitation; idiosyncratic works yield to socially shared musical conventions. Unlike the four-year-old, who sings a song of her own device, the seven-year-old will make musical gestures borrowed from tradition, which can be understood and appreciated by others. This change is not necessarily retrogressive; unless children acquire their heritage of shared musical ideas, their own works are likely to remain impoverished for lack of a sizable pool of materials to experiment with. Children have to acquire proficiency in established musical ideas before they can transcend them, first through speculative playfulness and later through metacognition. In speculative playfulness, an eleven-year-old may deliberately transform the commonplaces of the vernacular by, say, tonal inversion, experimenting with atonality, or ending a piece with a personalized tag. At about age fifteen, children enter the stage of metacognition; they begin to discover that certain kinds of music, one particular piece, or even a chord sequence corresponds with something they deeply feel (just as in the other arts they return again and again to a particular picture, a color scheme, or a line of poetry). The ultimate stage of metacognition is the budding young artist's effort, driven by a strong personal sense of values, to build systematically expressive universes of her own.[15] Few individuals in any but specialized (musical) societies reach this level.

At as early as thirteen months, children begin to scribble by vigorously and somewhat jerkily swinging the arm back and forth or up and down in arclike movements. Next come single arcs, which are steadily refined so that they become intentional horizontal or vertical straight lines. At about age three, youngsters begin to produce circles and squarish shapes amid the arcs and lines; these two-dimensional shapes represent objects for them. The shapes are added to and transformed so that even adults, without the child's explanation, can recognize them as humans, trees, flowers, houses, or animals. From ages five to eight, children use their rehearsed and perfected schemata to produce hundreds of pictures. In each, the focus tends to be on individual figures or items rather than on organic relationships among them or how they would appear to the eye in accordance with the laws of perspective. From seven to ten, children enter a phase of naturalistic representation. Animals are drawn with sufficient care to be recognizable as belonging to different species. Young artists are concerned to show the relative size of objects correctly, so that humans are no longer bigger than their houses, and they attempt perspective by adjusting the size of a tree to where it is placed, foreground or background.[16]

The first sources of joy for the very young scribbler are the learning of muscular control and of a skill and, concurrently, the use of that skill to produce tangible effects on the world. The aesthetic experience is as much a kinesthetic sense of mastery over the body, of the arm's rhythmic movement, as the feeling of satisfaction in the product on paper; or it may be that the tangle of arcs on paper is delightful to the very young child because it visibly confirms her kinesthetic competence. The achievement of straight lines, and then of circles and squares, further gives the child a sense of effective power to introduce order into the world. Pure design can appeal to children of all ages. Lines, shapes, and colors are symmetrically or rhythmically placed without any attempt at figural representation. The design itself pleases, in contrast to how Lewis remembered his own artistic efforts as a child. Even when pictorial subjects appear, the formality of the patterning suggests that the young artist strives for satisfaction in the picture's overall design rather than in the realism or expressiveness of the individual subjects.[17] Still, most children appear to be driven by the desire to tell a story, to present an event or a scene. They do so, however, without the techniques of mature art and without mature art's conventions and biases. They present the world they know in an exuberant and arbitrary manner, in disregard of perspective, proportion, relatedness of parts to the whole, and suitability of color. They work quickly, with little awareness of whether they are proceeding the right way, and they are pleased with the result, even though, after a while, a child may no longer recognize a work as her own.

Adults in Western society tend to find children's art at this stage refreshing and original and to see the next stage of pictorial realism as a lapse in creativity. However, this reaction is unjustified, for several reasons. First, in the earlier stage the young artist himself is quite unaware that he has done anything original: he is not fighting a convention because he does not know that any exists. Second, the young artist believes that his work, far from being subjective and arbitrary, presents a world readily entered into by others; for this reason, an adult's reiterated demands for explanation may trouble him. Third, even the fond parent should admit that out of the scores of pictures painted by her young child in perhaps a matter of weeks, only a few are truly worth pondering over for their boldness, unusual juxtapositions, and blends or clashes of color.

Children begin to aspire to pictorial realism as they increasingly recognize the existence of a common world. Participation in it calls for successful communication, which in turn requires clarity in presenting objective facts. They have now to take into account facts of perspective—what can and cannot be seen in a cluttered landscape, relationships of size, and changes in color with distance. Children at this stage tend to regard their own earlier artworks and the efforts of younger siblings as incompetent and arbitrary rather than as bold and creative. Ironically, at a time when the works of older children are judged conventional and aesthetically dull by adults, the children themselves show increasing sensitivity to the preeminent aesthetic values of style, expressiveness, balance, and composition. By the time they enter middle school, they also want to know how certain effects are achieved. This technical interest in "how" is driven, at least in part, by the recognition of beauty not only in art but also in the world at large.[18]

Growth in Sensibility

Even very young children can be enthralled by beauty. One way in which preschoolers and older children differ in the objects of their desire is scale. Preschoolers attend to individual objects; older children are able to go beyond the charm of concrete particulars to their composition. The young child's attention is caught by a tree stump or a boulder; the older child is able to appreciate an entire landscape, of which the tree stump and the boulder are elements. Preschoolers tend to be drawn to bright colors and to things that sparkle. Older children learn to appreciate subtler lights and hues in nature and in artifacts.

All aesthetic response must contain an element of magic; one is aware of being confronted by something out of the ordinary, miraculously right, "more real than real." But the worlds of young children and older ones are magical in different ways. Compare their responses to an illuminated Christmas tree—the younger child's ecstatic, the older one's cooler. A simple toy—crude from an adult's viewpoint—can seduce the young child. The magic is in the child's mind rather than in the toy, which serves as the launching pad for his own imaginative flights. As children grow older, they explore the object itself for aesthetic quality, which it must be seen to have to command renewed attention and respect.

The world has greater resonance for older children and adults than it does for the very young. Thus, to the teenager, a landscape can embody a mood. She looks at a landscape painting of dappled sunshine and picnic baskets and easily says of it that it is a *happy* scene. The younger child, by contrast, hesitates to attribute mood to inanimate objects and indeed even to human faces; he can be very matter-of-fact. Mood or atmosphere is an essential ingredient of the maturer person's sense of beauty, whereas this may not be so for the young child. Youngsters can enter a magical kingdom with ease; an up-turned chair becomes a fortress and the living room floor a battle-field. Older persons, too, can be affected by quite ordinary objects, but in a different—more contemplative and somewhat nostalgic—way. Thus, a tricycle abandoned on the sidewalk or a broken mirror in the drawer, which means little or nothing to a toddler, can have for older people poignancy—an aura of mortality and sadness that is not the work of unbridled imagination but appears to emanate from the thing itself.[19]

Experience, unless it carries resonance, is shallow and transient. Resonance is the result of the extension of one field of meaning to an-other—a change and enlargement of context so that a phenomenon is more than how it at first appears. What makes resonance possible is the human capacity for metaphorical perception and thought. Young children have it to a high degree. According to Howard Gardner and his coworkers, three-year-olds are already able to use a vivid, metaphorical language. They may, for example, compare a flashlight battery to a rolled-up sleeping bag, or nuns to penguins. However, the metaphors produced and appreciated by young chil-dren tend to depend either on perceptual resemblance (a pencil to a rocket ship) or on similarity of use; thus, a pencil used as a hairbrush is called a hairbrush. Expressive and psychological metaphors tend to be beyond their powers. Young children do not see a broken pencil as "sad" or a "loser." They find it difficult to understand that com-mon figure of speech—"heart of stone." They would not compare love to a summer's day.[20]

Young children may invent metaphors with wild abandon as though they were playthings to be tossed about. To an adult their analogies seem sometimes just right, but at other times wholly fan-tastic. When children are seven or eight, their production of meta-phor declines. They become more engrossed in facts and look down on their earlier verbal acrobatics as childish. They can still produce them on demand, but would rather not do so. As is already clear from their development in music and pictorial art, by the time chil-dren enter school, they seek to follow society's conventions and thereby enter a phase of realism. The mastering of the world's facts calls for close attentiveness. By paying close attention to the things out there, children learn to manipulate them and acquire the feeling of being in control. In the process, not only their calculating minds but also their imaginations reach outward and are strengthened. As they grow older, they are able to apprehend the world more fully and discriminately, discerning in it a subtleness and range of beauty, an expressiveness, an emotional and psychological depth that lie beyond the competence of the young child.

II

Sensory Delights

Pleasures of the Proximate Senses

*T*he senses, under the aegis and direction of the
mind, give us a world. Some are "proximate," others "distant." The
proximate senses yield the world closest to us, including our own
bodies. The position and movements of our bodies produce proprioception or kinesthesia, somatic awareness of the basic dimensions
of space. The other proximate senses are touch, sensitivity to
changes in temperature, taste, and smell. Hearing and sight are considered the senses that make the world "out there" truly accessible.
Since that distancing, momentary removal of self from object or
event is essential to the aesthetic experience, it is not surprising that
the aesthetic potential of the proximate senses has been undervalued

by Kant, among others. Yet the proximate senses, separately and together, add immensely to the vitality and beauty of the world, and distancing can and does occur in our experience of them.

Proprioception or Kinesthesia

"What is the proof of life? Movement," a ballet dancer writes. "And what higher and more beautiful movement is there than dancing? We use our bodies to manifest life itself."[1] Movement is indeed life. Most of us, however, are seldom conscious of producing beautiful movements, even though to a keen observer they may seem so. From early childhood onward we are taught, or learn through imitation, how to sit, stand, and move in appropriate, even graceful, ways. Learning such postures and movements—learning manners—is integral to the process of becoming part of one's culture. To feel self-conscious about movement is to risk awkwardness, insincerity (or at least the appearance of it), and immobilization. Although the ordinary movements and gestures of life often have a certain facility and flair, we tend to be aware of them only when they have become strained or inappropriate—that is, ugly. Movement is thus like health, usually taken for granted until there is some lack in it. Occasionally, though, just as there are times when we pause to savor our own physical well-being, so there are times when we know that we have made an exceptionally felicitous and efficient gesture.

Of course movement does improve with practice, but it is also a matter of talent. Some people can perform even the most ordinary tasks, like tying a shoelace or turning the pages of a book, with noticeable grace. Dancers and athletes, in particular, are aware of and enjoy their bodies' liberating power. Many of them discovered this awareness and enjoyment as children. Roger Bannister, who broke the barrier of the four-minute mile, recalls his exhilaration when, as a child, he ran "barefoot on firm dry sand." He remembers being "startled, and frightened, by the tremendous excitement that so few steps could create." "I glanced around uneasily to see if anyone was watching. A few more steps [and the] earth seemed almost to move with me. I was running now, and a fresh rhythm entered my body. . . . I had found a new source of power and beauty, a source I never dreamed existed."[2] These talented people may express exuberance kinesthetically on the spur of the moment, or recall their delight in bodily motion and contact with nature as young adults. One cold and crisp afternoon, the professional hockey player Eric Nesterinko drove his car unthinkingly onto a broad expanse of ice on the street. He got out of the car and put on his skates. "I took off my camel-hair coat. I was just in a suit jacket, on my skates. And I flew. Nobody was there. I was as free as a bird. . . . Incredible! It's beautiful!"[3] Albert Camus remembers vividly his youthful joustings with sea and sun in Algiers. "As I swim, my water-varnished arms flash out to turn gold in the sunlight, and then plunge back with a twist of all my muscles; the water streams along my whole body as my legs take tumultuous possession of the waves—and the horizon disappears."[4] For Camus, an event that occurred long ago, his sensuous joy in nature and in the competence of his own body—in movement—retains the immediacy of now, as though he had just climbed out of the water.

THE FLOW EXPERIENCE

Many people who do physical work well—carpentering, chopping down a tree, scything—feel visceral pleasure in the ease and naturalness of what they do, in their skill. Robert Hale is an academic anatomist. His delight in his work carries over into the domestic sphere. "You can't imagine," he says, "the silent fun I have at dinner parties while eating my food and dissecting a beautiful muscle. How I enjoy the spinalis dorsi—largely constituting a lamb chop. And I get almost sensual pleasure from skillfully dissecting a harmless pineapple, cutting all around the hard core."[5] Deep engagement in almost any kind of mental or physical activity can produce what the psychologist Mihaly Csikszentmihalyi calls the "flow" experience. "Flow" is not so much a scientist's technical term as how people in a wide range of occupations have described their feeling of apparently effortless control over what they do—a frictionless power the exercise of which generates a sense of being fully alive—fully aware of one's own movement and of one's environment.[6] A famous account of this flow experience in Tolstoy's *Anna Karenina* occurs in the midst of hard physical labor:

> Levin [the landowner] lost all count of time and had no idea whether it was late or early. A change began to come over his work which gave intense satisfaction. There were moments when he forgot what he was doing, when he mowed without effort and his line was almost as smooth and good as Titus's. . . . The longer Levin mowed, the oftener he experienced these moments of oblivion when it was not his arms which swung the scythe but the scythe seemed to mow of itself, a body full of life and consciousness of its own, and as though by magic, without a thought given to it, the work did itself regularly and carefully. These were the most blessed moments.[7]

EVERY MOTION A DANCE

In routine factory work, what counts is the tangible end product, not the workers' motions in its manufacture. Workers therefore tend not to see their bodies as instruments to be nurtured and trained. In sport, the goal is as precisely defined as in factory work: to reach a certain speed or height, to win against an opposing team. But unlike factory workers, athletes, to reach their goal, have to be highly conscious of the power and limits of their own bodies. The body is the athlete's instrument of success. It has to be nurtured and trained, mentally as well as physically. The athlete has to rehearse in his mind the necessary motions in relation to the barriers to be overcome. "A beautiful shot!" spectators exclaim as the ball rolls into the hole. The golfer himself senses the economy and elegance of the swing. In sport, success may be all-important, but the means to it have their own beauty and justification.

In dance, a particular gesture or motion is both a part of a larger composition and an end in itself: it must be artistically flawless. To keep their bodies and minds as attuned as possible to the aesthetics of stance and motion, some dancers train themselves to be continuously conscious of the poetry of movement in everything around them—a newspaper blowing down the sidewalk, a bird landing on the telephone wire—and in everything they do, like setting the table and sipping coffee. "Every day the whole day from the minute you get up is potentially a dance," says the choreographer and dancer Deborah Hay. "I dance by directing my consciousness to the movement of every cell in my body simultaneously so that I can feel parts of me from the inside out. . . . I dance by feeling the movement of space simultaneously all over my body so that it is like bringing my sensitivity to the very edges of my being from my head to my toe so that I can feel the movement of the air around me."[8]

Dancers can simultaneously project an image of total absorption and yet stand apart from themselves critically. They are able to view, Susan Foster writes, "their own bodies' effortless execution of the movement. Or they gaze up and out from the dance toward the audience, as if enjoining viewers to admire and take pleasure in their elegant rendition."[9]

Dance is the most ephemeral of arts. It is inscribed on air, not on paper, canvas, or stone. Except when captured by a movie or video camera, a work lasts no longer than the performance. Like composers, choreographers cannot expect their works to attain the degree of permanence—a reality to which artist and audience alike can return again and again—that writers, sculptors, and other artists in graphic media do as a matter of course. To the dancer, the end of a perfect line of movement marks the end of a beauty never to be precisely recaptured. The beauty of dance lies in part in this poignancy—an existence so fleeting that it seems, paradoxically, to transcend time.

Touch

The human skin is the most important human sensory system. As the anthropologist Ashley Montagu notes, "A human being can spend his life blind and deaf and completely lacking the senses of smell and taste, but he cannot survive at all without the functions performed by the skin."[10] Stimulation of the skin is necessary to the proper working of the digestive and eliminative organs, especially in the very young; it is essential to survival and growth. Mammalian mothers lick and groom their offspring for a considerable time after birth. In humans, according to Montagu, contractions of the uterus during the exceptionally long period of labor may provide necessary stimulation to fetal skin.[11] And all young mammals find pleasure in snuggling and cuddling up to another warm body.

That pleasure diminishes little with age. Initially it may seem to be a wholly inner-directed physical sensation, registered first at the skin surface and then quickly suffusing the whole body. We lower ourselves into a hot bath and register pure sensual delight from the temperature and feel of the water; later, drying off with a towel, we feel pleasure in the stimulated skin. But the mind often takes a part, too, appraising the water's qualities, appreciating the sensations of trunk and limbs soaking in the hot water and, later, the towel's fluffy, enveloping warmth.

"Tactile aesthetics," portentous and esoteric as it may sound, refers to the most common and necessary of aesthetic experiences. The pleasures of being alive and our deepest sense of well-being depend on cutaneous rewards that may come anytime, anywhere: the coolness of a stone in the shade, the warmth of a coffee cup, the smoothness of a baby's skin, the cuddling pressure of a heavy sweater, the silken texture of a kitten's fur, the roughness of a cobbled walk, the fat kiss of raindrops, "the delicious comfort of a balmy spring day as I walk beneath a row of trees and sense the alternating warmth and coolness of sun and shade,"[12] the feel of the carpet under the desk as you slip your stockinged foot out of your moccasin and brush your foot over its plush surface, and even "the slippage of the inner surface of the sock against the underside of your foot, something you normally only get to experience in the morning when you first pull the sock on."[13]

Modern society and scholars alike tend to discount the importance of tactile pleasures. Bodily contact and skin-to-skin stimulation, perhaps because they are thought to contribute merely to our physical well-being, are least esteemed in the scale of modern cultural values. "Have you hugged your child today?" This bumper-sticker tag suggests that even children cannot count on reassuring bodily contact. Among adults, touching and hugging as a manner of greeting are increasingly rare. In the United States, even the handshake is going out of fashion. The shower, the hot bath, and refinements such as the Jacuzzi have emerged as modern (not altogether satisfactory) substitutes for bodily contact; even in a society geared to "good times," one of humankind's greatest and most easily accessible sources of pleasure has become taboo.

We are also losing touch, in the literal sense, with nature. Children still enjoy jumping into a pile of leaves, slithering down a tree trunk, or rolling down a bank of snow. But adults have learned to enjoy nature mainly by simply looking at it. If they reach out for tactile rewards, they tend to do so in the context of strenuous sport: the impact of crampons on granite, the milky smoothness of ice beneath the skate, the violent rush of air in skydiving. D. H. Lawrence yearned for more direct and passionate contact, as his account of a man communing copulatively with vegetation shows:

> He was happy in the wet hillside, that was overgrown and obscure with bushes and flowers. He wanted to touch them all, to saturate himself with the touch of them all. He took off his clothes, and sat down naked among the primroses. . . .
>
> But they were too soft. He went through the long grass to a clump of young fir-trees, that were no higher than a man. The soft sharp boughs beat upon him . . . , threw little cold showers of drops on his belly, and beat his loins with their clusters of soft sharp needles. . . . To lie down and roll in the sticky, cool young hyacinths, to lie on one's belly and cover one's back with handfuls of fine wet grass . . . ; and then to sting one's thigh against the living dark bristles of fir-boughs . . . this was good, this was all very good, very satisfying.[14]

Like the other human senses, touch is exploratory and hence can open up a world. Lawrence's hero is discovering vegetation through the skin. But for most of us the human hand best embodies the seeking, searching, and appreciative nature of touch. Hands present us with a reality of discrete objects, many of which can be picked up and examined for their form, size, weight, and texture. Hands are restless; indeed, it is tempting to speak of them as curious. Children feel impelled to touch; for them, touch is a primary method of learning. In our society, young children are given soft, rounded things to play with. Among their early creations are mud pies; they knead and pat formless matter into just the right shape and, in the process, recognize the consistency and malleability of various kinds of earth. As they grow older, they play with harder and sharper objects; they learn the hardness of a ball from the way it bounces off a bat, the sharp corners of the "carpentered world" through playing with construction sets, the differing textures of nature through poking at them.

Touch is a delicate instrument for exploring and appreciating the world. No special skill is needed to feel the difference between a smooth pane of glass and one etched 1/2500 inch deep. Running a finger over bond paper, flower petal, and polished wood, we can tell that they are not the same, either in temperature or in texture. Training, naturally, increases sensitivity; professional "cloth feelers" in textile houses develop extraordinary competence in judging subtle differences.[15] But most of us have a comparable skill even without training. One example is our knack of telling small variations in the roughness of the sidewalk pavement by simply trailing a stick over it. Most surprising is the way we feel the texture, not at the area of contact between hand and stick, but at the end of the stick, as though it were an anatomical extension of ourself—what James Gibson calls an epicritical vibrissa.[16]

The better we are able to see and the more vivid and sharply detailed the world becomes, the safer and happier we tend to feel. A parallel relationship holds with touch, though we are not quite so aware of it, perhaps because pain, excesses of heat and cold, itchiness, and other irritations of skin are experiences that we can well do without: a certain numbing of awareness or anesthesia seems at times a blessed relief. Yet to be deprived of the sensitivity and flexibility of the hand, even briefly and in small measure, is frustrating, as we realize when we drive a car with heavy gloves on or try to tie a shoelace with frozen fingers. Although young children are tolerant of jam smeared on their face and hands, most adults find stickiness a strangely uncomfortable sensation. We should like to *feel* the beauty of the world. But the more we are aware of the world's abundant offering of tactile and thermal delights, the more we are also aware of its repellent aspects: the tactile pleasures of living grass and human hair are countered by the disgust (partly but perhaps not wholly induced by culture) with which we respond to contact with unhealthy skin, feces, or a corpse.[17]

LANDSCAPES OF TOUCH

Most tactile sensations reach us indirectly, through the eyes. Our physical environment feels ineluctably tactile even though we touch only a small part of it. Reddish fluffy surfaces are warm, light-blue glittering ones cool. A glass coffee table next to a polished walnut chest is a tactile composition. A street lined by brownstone houses and graceful trees makes a charming picture, but the charm comes as much through our sense of touch as through the eyes. Seeing and the tactile sensation are so closely wed that even when we are looking at a painting it is not clear that we are attending solely to its visual qualities. Bernard Berenson surprised the art world in 1896 by emphasizing that a painting must possess more than just visual excellence; it must have "tactile values" that reach out and touch, even embrace, the viewer.[18] More recently, the art critic Robert Hughes has written of John Constable: "His childhood was substance rather than fantasy: tactile memories of mold, mud, woodgrain and brick became some of the most 'painterly' painting in the history of art. The foreground of *The Leaping Horse* is all matter, and the things in it— squidgy earth, tangled weeds and wild flowers, prickle of light on the dark skin of water sliding over a hidden edge—are troweled and spattered on with ecstatic gusto. This is the landscape of touch."[19]

One reason for the strong appeal of nature is the range and complexity of its tactile impress. In a small nook of the mineral world, one may find granite and sand, gnarled lava and viscid mud. A single flowering plant may have rough bark, smooth waxy leaves, and satiny petals. That wonderful work of nature, the human body, is exceptionally rewarding as an object for tactile exploration: an experienced lover's hand moves from rumpled hair and smooth firm breast to the soft skin below the armpits, the muscled thighs, the knee's hard knob, and registers at the same time a surprising range of temperature from the cool tip of the nose to the heat of the groin.

Landscape designers try to imitate, and improve on, nature's sensual wealth. Asian gardens are created with visual-tactile qualities in mind. A Chinese garden is composed of *yin* (soft) and *yang* (hard) elements—the softness of water and of undulating perforated garden walls, the hardness of craggy limestone rocks. An Islamic garden is a concordance of sight, sound, and scent, but it is also an oasis of thermal delight—its shades and coolness contrasting vividly with the glare and heat beyond its high walls. Similarly, modern householders more or less consciously arrange the interiors of their homes so as to provide comfort and variety to the sense of touch: a long-stemmed aluminum lamp may be placed next to heavy curtains, a table on spindly legs next to a leather ottoman, a bearskin rug on a polished parquet floor. Urban places, too, appeal to more than the eyes. An old European town with cobbled streets and half-timbered houses opening onto a sun-drenched plaza is a visual-tactile feast.[20] It was not necessarily planned with such aesthetic effects in mind; these emerged as the result of happy happenstance and, more important, sensibilities that have been tested and refined over the centuries. We all know of cities or parts of cities, both old and new, that feel eminently livable, although why they do so eludes analysis. A secret of their success may lie in their tactile variation and warmth.

Like all other senses, the tactile sense is activated by contrast—alternations of heat and cold, roughness and smoothness, lightness and weight. Perhaps the range of appreciation (or tolerance) is narrower than that for, say, sight because it is more intimately bound to the basic physiological processes, to elemental moods and emotions.

Touch is the sense least susceptible to deception and hence the one in which we tend to put the most trust. For doubting Thomas, seeing was not believing; he had to touch the resurrected Christ to believe. The real, ultimately, is that which offers resistance. The tactile sense comes up *against* an object, and that direct contact, felt sometimes as harsh impingement, is our final guarantee of the real.

What is the relation between reality and beauty, and of both to touch? Both beauty and reality are governed by necessity; both project a sense of the inevitable that transcends mere human will and desire. Beauty is "the order of the world" (Simone Weil), manifest in mathematics, where lawful and necessary relations rule; and it is manifest also in the forces of nature—in gravity, which has impressed folds upon the mountains and on the waves of the sea.[21] There is nothing arbitrary about these folds. They exist in obedience to natural forces, and a source of their beauty lies in our recognition of that fact.

The order of the world is accessible to us indirectly, says Weil, through the image. By means of the image, itself wholly dependent on an act of close attention, "we can contemplate the necessity which is the substance of the universe." But necessity can be known directly only through physical contact—"by the blows it deals." Physical labor is, for Weil, direct contact with necessity and the real, and hence the beauty of the world. Moreover, in its best moments, physical labor "is a contact so full that no equivalent can be found elsewhere." Artists, scientists, and contemplatives, trying to penetrate the veils of fantasy to reach the real, may instead add to the unreal by fabricating their own illusions. Those who earn a living through physical labor are less likely to fall under illusion, for they feel the impingements of matter almost constantly. "He who is aching in every limb, worn out by the effort of a day of work, that is to say a day when he has been subject to matter," bears in his flesh the reality and beauty of the universe.[22]

But we not only are impinged upon by external reality; we also impinge—that is, exert force—on it. Touch, unlike the other senses, modifies its object. It reminds us that we are not only observers of the world but actors in it. With this awareness comes pride in our ability to do and make, but a pride that is shadowed by guilt, for unmaking precedes making: we are both destroyers and creators.

Eating, Taste, and Culture

Eating is a mode of touch. "Eating is touch carried to the bitter end," says Samuel Butler. It forcefully reminds us of our animal nature. Culture masks human animality; when the mask slips, the fact that we live by devouring other organisms rises to haunt us.

Watching people eat and noting what they eat, especially if they are of a different culture, is seldom an elevating experience. For the Chinese, eating has close ties to health, medicine, and a cosmological world view. Food preparation and consumption are considered an art. Yet these elevated cultural concepts are not always evident, at least not everywhere in China, and not to an outsider. Colin Thubron, for one, finds the Chinese national obsession with food, ascending to "a guzzling crescendo," repellent. He describes an eating quarter in the nontourist part of Canton around 1980. Feasters, mostly men, gather around the tables. "Every course drops into a gloating circumference of famished stares and rapt cries. Diners burp and smack their lips in hoggish celebration." The concept of taboo, Thubron notes, seems wholly absent. "In Cantonese cooking, nothing edible is sacred. It reflects an old Chinese mercilessness toward their surroundings. Every part of every animal—pig stomach, lynx breast, whole bamboo rats and salamanders—is consumed." Thubron, searching for something he can eat, enters a rowdy restaurant, where the waitress relentlessly plies him with "shredded cat thick soup and braised python with mushrooms."[23]

This account, blind though it is to different tastes and values, should make any thinking person uneasy about food preparation and eating. In these activities, biological imperatives are worrisomely joined to sensual delight, the killing and evisceration of living things to art, animality to the claims of culture, taste (a process in the mouth's cavern) to that refined achievement known as "good taste." As people become more and more conscious of their status as dignified cultural beings, eating/tasting tends to be done in public only when it is accompanied by some other, more obviously respectable activity, such as social conversation and music. And if one has to eat alone in a restaurant, one pretends to be engaged simultaneously in the higher occupation of reading a magazine.

Nevertheless, eating/tasting is a cultural activity in its own right, with a long history. Let us consider the aesthetics of food preparation and eating in two parts of the world—Europe and China.

FOOD AND MANNERS IN EUROPE

From the Middle Ages to modern times, Europe underwent progressive refinement in all things concerning food—its preparation, the dishes offered, table manners, utensils, and the larger setting of hall or room. It experienced the development of taste—good taste. The progression was by no means linear. There have been several swings between an ideal of simplicity and an ideal of luxuriance, between foods commended for their natural and exquisite flavor and foods commended for their symphonic richness—the effect of using artfully simple ingredients.

In the late Middle Ages, food still tended to be messy, prepared pell-mell. Expensive spices and viands might find themselves bedded in the same dish with meats that were none too fresh and were very probably contaminated by the filth in the kitchen. The range and quantity of food would have bewildered and repelled a modern gourmet. Plantagenet kings of the fourteenth and fifteenth centuries ate everything that had wings, from bustards to sparrows, herons, egrets, and bitterns; and everything that swam, from minnows to porpoises. Medieval cooks used vegetables and herbs profusely and indiscriminately. Many dishes were created by combining every scrap of greenstuff that came to hand. In a hare stew one might find cabbage, beets, borage, mallows, parsley, betony, the white part of leeks, the tops of young nettles, and violets. Roses, hawthorn, and primroses might also find their way into a dish.[24]

For the wealthy, a medieval dinner consisted of two or three courses, but each course could contain more than a dozen different kinds of food heaped high on large platters. Guests were confronted by such rich fare as shields of boiled and pickled boar, hulled wheat boiled in milk and venison, oily stews, salted hart, pheasant, swan, capons, lampreys, perch, rabbit, mutton, baked custard, and tart fruit. The second course was again made up of a large array of rich meat and fish hardly distinguishable from the first.[25] The concept of an orderly sequence—soup, fish, meat, and dessert—did not appear until the end of the seventeenth century. Copiousness, rather than discrimination, was the key concept in premodern culinary art. Cooks were indifferent to the unique textures and flavors of the materials that went into a cauldron. The French critic and poet Nicolas Boileau-Despréaux mentions an enormous mixed grill consisting of a hare, six chickens, three rabbits, and six pigeons, all served on the same plate. These hodgepodges were relentlessly overcooked, probably because the game was usually "high." One of the recipes for ragout, which Louis XIV and his courtiers were fond of, called for putting a number of different kinds of flesh and fowl in a cauldron, adding a large quantity of spice, and stewing the mixture for twelve hours. It is unlikely, in W. H. Lewis's view, that this dish "would be saved at the twelfth hour by a lavish top dressing of musk, amber, and assorted perfumes."[26]

Before the seventeenth century, the preparation of most dishes, even of the pâtés, meats, and side dishes of ostentatious feasts, required little imaginative forethought. Thereafter, carefully prepared meals for discerning people began to emerge. The French words *gourmand* and *gourmet*, both initially used to express unqualified approval, won general acceptance in urbane Europe. A further sign of refinement in taste lay in the serving of foods on several small dishes rather than on a few large platters. Incompatible flavors were thus kept apart. After 1700, more and more diners accepted the notion that the distinctive flavor and texture of a dish rather than the quantity and expense of its ingredients should be the primary criteria of excellence. The care that went into cooking by the middle of the eighteenth century is suggested by the menu for a reception in honor of the Archbishop of Besançon. Among the dishes listed were "Bisque d'écrevisse, potage à la reine, grenouilles à la poulette, truites grillées, anguilles en serpentin, filets de brochet, carpes du Doubs avec coulis d'écrevisse, tourte de laitances de carpes."[27]

After the French Revolution, France led Europe in transforming cooking into an art in the grand style and an honored profession, with its own literature and roster of famous names. The most distinguished chef of this time was Antonin Carême. In the creation of dishes he strove, paradoxically, for both ostentation and simplicity. Trained in *pâtisserie*, an art that encouraged creative leanings, he extended the architectural style to cooking generally. For a grand dinner, he might erect picturesque ruins made of lard and Greek temples in sugar and marzipan so that the gastronome's mind, and not just his palate, could be pleasurably stimulated. Carême's creations were also architectural in that they had a "built" character: they were made of purées, essences, and sauces that were themselves complex creations and yet were listed simply as ingredients along with a piece of celery or a chopped onion. A dish, in other words, was the culmination of a long and elaborate process.[28]

Carême achieved simplification by eliminating medieval survivals such as trimmings of cockscombs and sweetbreads. More important, he established the principle of garnishing meat with meat, fish with fish. His culinary aesthetic is caught in Lady Sydney Morgan's description of a dinner at the Baron de Rothschild's: "no dark-brown gravies, no flavour of cayenne and allspice, no tincture of catsup and walnut pickle, no visible agency of those vulgar elements of cooking of the good old times, fire and water. Distillations of the most delicate viands, extracted in silver dews, with chemical precision . . . formed the *fond* of all. Every meat presented its own natural aroma—every vegetable its own shade of verdure."[29] Even in Carême's elaborate achievements, his aim was not to superimpose and confuse flavors, but rather to isolate and throw them into relief.[30]

Despite his own success in creating "simple" and distinctive flavors, in general Carême's approach encouraged ostentation and, with it, the sacrifice of savor for grand visual effects. Master cooks had yielded to this temptation since at least Roman times: thus Petronius described a feast in which a hare was tricked out with wings to look like a Pegasus, and roast pork carved into models of fish, songbirds, and a goose.[31] For millennia, then, chefs in the West have often "played with food," treating edibles as materials for sculpture and architecture, as though the creation of alluring flavors could not in itself win for them high standing.

Closer to our time, Georges Auguste Escoffier rivaled Carême in celebrity and influence. Like his predecessor, Escoffier was capable of architectural grandeur, but his reputation rested even more on achieving a perfect balance between a few superb ingredients—sometimes such rare items as truffle and crayfish, but also quite ordinary ingredients that even the most common middle-class kitchen could afford. One of Escoffier's best-known creations was Peach Melba: to a coupling of vanilla ice cream with peach, he gave a final touch of perfection by balancing the smooth sweetness of the cream and the textural resistance and flavor of the peach with the tartness of raspberries.[32]

Food is basic, and people's taste for food tends to be traditional, conservative, associated with old family recipes and perhaps also regional ones. People tend to like what they have always had. The business of "inventing" new flavors seems a questionable venture. Yet cooking and the tastes developed for it have a long history of innovation and change. In post-Revolutionary France, chefs who established their own businesses after the departure or demise of their aristocratic patrons competed with one another for a growing, gastronomically sophisticated clientele. They were driven to offer fresh gustatory pleasures. In the nineteenth century, this demand for originality became ever more insistent. Escoffier, to maintain his reputation, was compelled to devise new dishes all the time. He writes: "I have ceased counting the nights spent in the attempt to discover new combinations, when, completely broken with the fatigue of the heavy day, my body ought to have been at rest."[33]

The greatest challenge lay not in the profusion of expensive ingredients and strong flavors, but (as Escoffier had seen) in obtaining a deeply satisfying gastronomic experience with a few choice ingredients. In the history of Western cooking, the virtues of simplicity and subtlety were periodically recognized and elevated to serve as criteria of excellence. The latest rejection of rich sauces and compli-cated foods in general occurred in the 1960s, giving rise to a style known as *nouvelle cuisine*. Chefs insisted on buying the freshest veg-etables and meats available in the market each day. The elevation of freshness—the desire to bring out the qualities inherent in the ma-terial—called for reduced cooking time for most seafoods, game birds, and veal, but especially for green vegetables. Steaming as a method of cooking also found favor in *nouvelle cuisine*. Both the em-phasis on reduced time and on steaming reflect Chinese influence.

FOOD AND MANNERS IN CHINA

Perhaps no other civilization has put as much emphasis on the art of cooking or taken so much pleasure in food as the Chinese. Since the earliest times, cooking in China has carried a prestige unmatched and perhaps somewhat incomprehensible to people in other cul-tures. The *Li Chi*, a Confucian classic with materials dating back to the fifth century B.C. and earlier, treats the evolution of culture as though it were a matter of the evolution of cooking skills.

> Formerly the ancient kings knew not yet the transforming power of fire, but ate the fruits of plants and trees, and the flesh of birds and beasts, drinking their blood and swallowing the hair and feathers. The later sages then arose, and men learned to take ad-vantage of fire. They toasted, grilled, boiled, and roasted. They produced must and sauces. . . . They were thus able to nourish the living, and to make offerings to the dead; to serve the spirits of the departed and God.[34]

Ritual disciplines attention and encourages people to develop their powers of discernment and discrimination. "In putting down a boiled fish to be eaten," the *Li Chi* asseverates, "the tail was laid in front. In winter it was placed with the fat belly on the right; in sum-mer with the back. . . . All condiments were taken up with the right hand, and were therefore placed on the left. . . ."[35] Every-thing in ritual, not least food, must be done correctly. Confucius re-portedly "did not eat meat which was not cut properly, nor what was served without its proper sauce."[36] Clearly, the sage was a fastidious ritualist, but this fact does not exclude the likelihood that he also had a sophisticated palate and a well-developed aesthetic sensibility.

Food is also medicine, and no doubt the understanding of foods—not only their texture and taste but their nutritive and curative powers—has gained enormously by the association. In China, the exceptionally varied kinds of food eaten reflect in part the poverty of the people, who could not afford to disdain anything that assuaged hunger, and in part the unending search for *materia medica*—the healing qualities of plants, animals, and their minutely differentiated parts. That health depends on dietary regulation is a belief that the Chinese have shared with Westerners through the millennia. What distinguishes the Chinese is the way they have subsumed food and medicine under the overarching, universal principles of *yin* and *yang*. Most foods can be classified as having either *yin* or *yang* properties, and the wise eater is one whose diet exhibits a proper balance. Harmony in food is the desideratum, as it is in all other areas of Chinese life. Extremes and excess are to be avoided.

Nevertheless, excesses have occurred. Ordinary people tended to overeat because they had no assurance that food would always be available to them. The rich offered mountains of food on social occasions and overindulged from the desire to impress. An eighteenth-century poet and hedonist, Yuan Mei, wrote in his book *Recipes from Sui Garden*: "I always say that chicken, pork, fish and duck are the original geniuses of the board, each with a flavour of its own, each with its distinctive style; whereas sea-slug and swallows-nest (despite their costliness) are commonplace fellows, with no character—in fact, mere hangers-on." But swallows-nest carried prestige, whereas mere chicken and pork did not. When a provincial governor offered Yuan Mei "plain boiled swallows-nest, served in enormous vases, like flower-pots," the poet was unimpressed and declared that "it had no taste at all."[37]

The other extreme is abstinence. In the Western world, hermits and other aspirants to spiritual elevation have restricted themselves to stale bread and water. The Chinese have seldom carried abstinence so far. Buddhists, required to abstain from meat, ingeniously concocted "imitation meats" of the most mouth-watering texture and flavor. A scholar-official, disaffected with the luxuries of city and court, might sing of rural simplicity, but it was a simplicity that— as least in matters of food—did not necessarily sacrifice taste. Thus a thirteenth-century dramatist envisaged, no doubt somewhat ideally, a simple harvest meal under the gourd trellis, where workers could "drink wine from earthen bowls and porcelain pots, swallow the tender eggplants with their skins, and gulp down the little melons, seeds and all."[38]

The Chinese love of food—their search for pleasures of the palate—is reflected, first, in the sheer variety of what they eat. The Chinese are true omnivores, with few taboos, and these local. Frederick Mote writes of the animal fare alone: "Beyond such relatively ordinary items (in the West) as hares, quails, squabs, and pheasants, a Ming dynasty (1368–1644) source also mentions as standard foods: cormorants, owls, storks and donkeys, mules, tigers, deer of several varieties, wild boars, camels, bears, wild goats, foxes and wolves, several kinds of rodents, and mollusks and shellfish of many kinds."[39]

A second indicator of this love of food is the number and popularity of specialized restaurants. In Hang-chou, at about the time Marco Polo visited it, a resident could go to a place where only iced food was served; other specialized shops and their offerings included "the sweet soya soup at the Mixed-Wares Market, a pig cooked in ashes in front of the Longevity-and-Compassion Palace, the fish soup of Mother Sung . . . , boiled pork from Wei-the-Big-Knife at the Cat Bridge, and honey fritters from Chou-number-five in front of the Five-span Pavilion. Among the more exotic dishes were scented shellfish cooked in rice wine, goose with apricots, lotus seed soup, pimento soup with mussels, fish cooked with plums; and among the most common, fritters and thinly-sliced soufflés, ravioli and pies."[40]

A third indicator is the exceptionally rich vocabulary of taste and texture. The words used are not limited to cooks and gourmets but are a part of ordinary speech in daily use. *Ts'ui* describes a highly desired texture closely tied to freshness and the critical importance of not overcooking. *Ts'ui* offers resistance to the teeth followed by, as E. N. Anderson puts it, "a burst of succulence," as exemplified in newly picked bamboo shoots, fruit ripe enough to eat but not soft, fresh vegetables quickly stir-fried, and chicken boiled a very short time so that it is just done. Anderson goes on to list other evaluative words: "*shuang* (resilient, springy, somewhere between crunchy and rubbery, like some seaweeds), and *kan* (translated 'sweet,' but including anything with a sapid, alluring taste). Fried foods should be *su*—oily but light and not soggy—rather than *ni* (greasy). Above all, foods should taste *hsien*, which means not just fresh but *au point* in general. . . . In south and east China particularly, foods are often praised by being described as *ch'ing*, 'clear' or 'pure.' This means that they have a delicate, subtle, exquisite flavor—not obtrusive, heavy, or harsh."[41]

The final reflection of the Chinese love of food is the detailed knowledge of its geography, possessed by the literati simply as part of the baggage of being cultured. A special food is often named after its locality—for example, Peking duck. The way to offer an irresistible invitation to dine is to say that a delicacy—be it only a vinegar—has been obtained from a locality famed for that particular product. Connoisseurship of tea and spring water can be carried to great heights. A scholar of the early seventeenth century, Chang Tai, was such a connoisseur. One day he called on a fellow scholar and expert on tea, Min Wenshui, who lived in another town. As soon as he entered Min Wenshui's residence, his nostrils were assailed by a wonderful fragrance.

> "What is this tea?" I asked. "Langwan," Wenshui replied. I tasted it again and said, "Now don't deceive me. The method of preparation is Langwan, but the tea leaves are not Langwan." "What is it then?" asked Wenshui smiling. I tasted it again and said, "Why is it so much like Lochieh tea?" Wenshui was quite struck by my answer and said, "Marvelous! Marvelous!" "What water is it?" I asked. "Huich'uan," he said. "Don't make fun of me," I said. "How can Huich'uan water be carried here over a long distance, and after the shaking on the way, still retain its keenness?" So Wenshui said, "When I take Huich'uan water, I dig a

well, and wait at night until the new current comes, and then take it up. I put a lot of mountain rocks at the bottom of the jar, and during the voyage I permit only sailing with the wind, but no rowing. Hence the water keeps its edge."[42]

To the modern Western reader, this exchange between two Chinese friends on the quality and provenance of tea and water may seem excessive—an exercise in connoisseurship better suited to the high arts of music and painting. The exchange does show, however, the extraordinary importance the Chinese have traditionally given to the palate. To be cultured is, first, to know the rites and the classics; second, to have a certain flair for poetry and painting; third, to be an aesthete of food—to appreciate its precise flavor and texture. But although good professional cooks are respected, cooking, with its unavoidable violence of chopping and cutting, boiling and frying, and its intimate association with blood and death, tends to arouse unease. Tasting itself should be suspect, since it ends in destruction, but somehow the Chinese (like people in the West) have managed to repress this knowledge. Tasting now seems almost wholly an aesthetic activity.

Smell

When we say that a food tastes good, we are not speaking with physiological precision, for much of what is tasteful in food is contributed by smell. The organs of taste and smell are so close to each other and their effects so thoroughly interfused that we habitually treat them as one.

Smell gives us many innocent pleasures, yet its aesthetic standing, particularly in the Western world, is low. Kant denied aesthetic standing to smell altogether.[43] Freud attributed the "fateful process of civilization" to "man's adoption of an erect posture," which had the effect of giving the eyes rather than the nose the primary role in sexual excitation.[44] In the eighteenth century, European travelers and thinkers believed that the sense of smell was better developed among savages, especially young ones, than among the civilized.

People of refinement took for granted that smell lay at the bottom of the hierarchy of senses; they were suspicious of sniffing and smelling and of any predilection for powerful animal or sexual odors. It was and still is a common prejudice to regard the eyes and the ears as the noble organs, and the nose as primitive or animal.

BONDING AND MEMORY

Humans *are* primarily visual beings. Our distance sensor is the eye, unlike nonprimate mammals, for whom the nose opens up space and pinpoints distant objects of desire. The sensitive surface of a dog's or a deer's nose is some thirty times larger than it is in a human; their olfactory world is not only larger but filled with a degree of detail and delicate discrimination that people can barely imagine.[45] Smell, compared with sight and hearing, affects our emotions at a more deeply buried level. The olfactory sense is linked to a primitive part of the brain that controls emotions and mood and the involuntary movements of life, including breathing, heartbeat, pupil size, and genital erection. The smell of food can make one's mouth water; the musky odor of sex—the yeasty "baked bread" fragrance of the body—can cause involuntary sexual arousal. Somerset Maugham's test for whether a woman is truly in love with a man is not whether she likes how he looks but whether she likes the smell of his raincoat. The appeal of odor goes beyond, or below, reason. Polite society will not admit that foot fetishism is as much olfactory as visual bondage. A dedicated few regard even the armpit as a "charming grotto, full of intriguing odorous surprises."[46]

A mother who takes time to admire the delicate and perfectly formed fingers of her infant is having an aesthetic experience—one of many such experiences that strengthen the mother-and-child bond. Odor cements the bond, too, but at a less conscious level. At only six weeks an infant shows a strong preference for its own mother's scent over that of another woman. And the mother is able to distinguish by scent alone her two-day-old infant's bassinet from another's. A child's own earliest attachment to environment may well be acquired through the nose. The mother is the newborn's first and most significant place, recognized more by scent than by facial configuration. When the child is old enough to walk and explore its nonhuman environment, it clutches a blanket as a source of comfort and security. The soft tactile quality of the blanket gives reassurance; no less comforting, however, is its rich mix of human odors.[47]

The directness and immediacy of smell provide a sharp contrast to the abstractive and compositional proclivities of sight. Perhaps for this reason an odor can resurrect the past with a vividness that no visual image can. I have direct evidence of this phenomenon. I returned to Sydney, Australia, twenty-three years after leaving it as a child, expecting the experience to be filled with nostalgia. This did not happen. Sydney had been transformed during my absence; many old landmarks had been demolished and supplanted by gleaming new buildings. My old home was still there, next to a beach, bordered by a promenade. The beach, the promenade, and a little playground with its row of swings remained much the same. Still, the past eluded me. I could not project myself back into childhood except as an intellectual exercise. Although the physical place had not altered, my perception of it had. I saw the beach one way as a child; as an adult, I saw it in quite another way, with different focuses and values. My eyes failed me in my quest. But my nose did not, for just as I was about to conclude that I could not go home again, a strong whiff of seaweed assaulted my nostrils, and I was thrown back to childhood. For a fleeting instant I stood on the beach, a twelve-year-old again. Odor has this power to restore the past because, unlike the visual image, it is an encapsulated experience that has been left largely uninterpreted and undeveloped.

SAVOR OF LIFE

Until they are deprived of their sense of smell, people tend to be unaware of the fact that odor contributes much of the savor of life. A common cold curtails smell for a short time. A head injury, by contrast, can so damage the olfactory tracts that the victim permanently loses the world of odor. The loss is keenly felt. One man who suffered such a debilitating injury says: "You *smell* people, you *smell* books, you *smell* the city, you *smell* the spring—maybe not consciously, but as a rich unconscious background to everything else. [After the accident] my whole world was suddenly radically poorer."[48] Food without aroma has little appeal; indeed, it can be repellent. One then eats simply to survive. Another victim of such a loss says: "Of course, I had to eat—I didn't want to die. But the regular food—it was all like garbage. . . . I could drink a little cold milk. I could eat a little cold boiled potato. . . . I could eat a little vanilla ice cream. That stuff, it didn't taste good, but it didn't taste bad. It didn't have any taste at all." This victim can no longer bear to drink coffee; the memory of its aroma, now only a memory, is intolerable.[49]

The loss of a taste for food is often accompanied by a loss of appetite for sex. Robbed of scent, life and the world become gray and passionless. Visual and auditory beauties cannot altogether compensate for the deprivation of stimuli that arouse the deepest emotions and instincts of one's animal nature.

What if one possessed an extraordinary sense of smell? What if smell were the dominant sense, as it is in a dog? Of course, we cannot fully enter another species' experiential reality, but we can gain an insight into it from abnormal experiences, such as the one reported by the neurologist Oliver Sacks. A medical student, when he was on amphetamines, dreamed vividly one night that he had become a dog living in a world of rich odors. Waking, he continued to find himself in just such a world.

> He experienced a certain impulse to sniff. . . . Sexual smells were exciting and increased—but no more so, he felt, than food smells and other smells. Smell pleasure was intense—smell displeasure, too—but it seemed to him less a world of mere pleasure and displeasure than a whole aesthetic, a whole judgment, a whole new significance, which surrounded him. "It was a world overwhelmingly concrete, of particulars," he said, "a world overwhelming in immediacy, in immediate significance." Somewhat intellectual before, and inclined to reflection and abstraction, he now found thought, abstraction and categorisation, somewhat difficult and unreal.[50]

AN EDUCATED NOSE

From one angle, we may view odor as "primitive," something intimately associated with food and sex. From another angle, we can see that the discernment and appreciation of fragrance are capable of endless refinement. In humans, the sense of smell, no less than the other senses, must be developed under the aegis and pressure of culture if it is to fulfill its potential. Young children may have a keen sense of smell, but their olfactory world is limited by their narrow bands of attention. They crawl on the floor, which is a rich source of odor, or walk through meadows and grassy fields, where the plants' aromatic crowns may rise to the height of their faces; but unless they attend actively the odors escape them, or take hold only at an unconscious level never to be retrieved except perhaps through accidental jolts of memory.

The human sense of smell begins to be operative early. Even a couple of days after birth, newborn babies show sensitivity to odors. However, the response is to intensity rather than to character or type; there is little evidence that they appreciate one odor over another. A toddler who likes safrol more than butyric acid on one trial may well reverse his opinion on another trial. Preference becomes somewhat less arbitrary as children grow older. From age four onward they show an increasing tendency to discriminate and differentiate between pleasant and unpleasant odors, and their preferences move from fruity fragrances to those of flowers and, still later, to fragrances of greater complexity and subtlety.[51]

Despite the endearing myth that Ferdinand the Bull loves flowers, there is little evidence to show that bulls enjoy floral fragrance. Dogs, too, for all the sensitivity of their noses, seem little aware of the olfactory appeal of fruits and flowers. Yet liking for floral fragrance is widespread among humans in different parts of the world. This preference serves no discernible adaptive purpose; it is largely cultural. "The perfume of the honeysuckle and the sweet mellifluous scent of the countryside in the May sun are doubtless of practical use to the bee," says R. W. Moncrieff, "but to the human their value is mainly aesthetic and [somewhat] abstract," and perhaps also enhancing of "emotional maturity."[52]

FRAGRANCES IN NATURE

Although the olfactory sense can give us not only physiological stimulation but a world of odors as aesthetically satisfying in its way as the worlds of sight and sound, it is not cultivated to anything like a similar extent. Pictorial geographies fill libraries, as do essays and poems praising the visual diversity and splendor of nature. But an olfactory geography and aesthetic is practically nonexistent. Few cultures acknowledge odor as a principal component of the natural environment. The introduction of fragrance into one's setting is widely taken to be, at most, a supplementary or minor art. In China, for instance, numerous poems laud nature's shapes, colors, and even sounds. "Homing birds head for the tall trees, / p'ien-p'ien go their swift wings flapping." The use of onomatopoeia is commonplace, but references to odor are rare even where we would expect them: "Summer plums—crimson fruit chilled in water; / autumn lotus root—tender threads to pluck; / but our happy days, when will they come?"[53] Despite the power of odor to recapture the past, Chinese poets, for all their desire to evoke nostalgia, seem insensitive to its magic. When they do use fragrance in a poem, they almost always associate it with the presence of the human female; but nothing in Chinese literature matches the extended use of scent to embody the intensity and richness of love as in King Solomon's Song of Songs.

Some kind of emotional bond with nature is universal. Aesthetic appreciation, including pleasure in nature's odors, is, however, far less common, for such appreciation presupposes a large measure of confidence in relation to nature. In the Western world, wilderness elicited delight only when it no longer overwhelmed. Mountains and vast forests gained in aesthetic value as they became more accessible and lost their reputation for harboring hostile animals, humans, and demons. This process was slow and uneven. As late as the eighteenth century, the age-old fear of nature had not totally disappeared even among well-educated Europeans. Odorous effluvia from the earth contributed to the fear. "The air of a place was a frightening mixture of the smoke, sulfurs, and aqueous, volatile, oily, and saline vapors that the earth gave off." Quarries were open sores, exuding terrible threats from the depths of the earth—"metallic vapors" that attacked the nostrils and the brain. Peasants in the countryside were unhealthy because, when they bent down to till, they brought their faces too close to the soil. Cultivation, especially of virgin land, exposed whole villages to the threat of "morbific vapors." Water in almost all forms—fog, dew, and airs wafted in from the sea—was suspect even apart from the smell, but without doubt the greatest danger was perceived to come from the marshes and swamps, where vegetation and animal corpses fermented and decayed, letting lose a vile-smelling toxic miasma.[54]

As in Europe, appreciation of nature emerged in China only as people became less suspicious of it. When northerners from the heavily settled and flattish Huang Ho basin migrated southward into a mountainous land covered with dense tropical forests, they were ambivalent about the pungent odors they encountered there. Poets of the T'ang dynasty distinguishing between the relatively odorless North and the redolent South seldom took notice of the fragrance of exotics like rosewood, camphor, or cloves; rather, they were drawn to the familiar scents of the flowering orange and tangerine.

> *The forest is darkened by interlaced liquidambar leaves,*
> *The gardens are fragrant from tangerine blossoms.*
> *But who is my neighbor here beyond the waste?*
> *To console my loneliness clouds and sunsets must suffice.*

Fragrance evoked mild pleasure and wistful melancholy rather than joy. Edward Schafer observes that although the pleasant aromas of the South, "so near to the sweet-smelling gardens and temples of Indianized lands," could have inspired comparisons with the Perfumed Land of the Buddha, this did not happen. To the Chinese, "here was no Eden, but a fearful wilderness, whose evils were only partly offset by the awareness of orange blossoms in the heavy air."[55]

FRAGRANCES IN THE COUNTRYSIDE

The small literature that exists on the smell of humanized landscapes (other than flower gardens) focuses overwhelmingly on bad or mixed odors. In the eighteenth and nineteenth centuries, Europeans were acutely conscious of the odors of decay and death in their built environments, particularly in prisons, asylums, workhouses, and in the quarters of the poor, but also in the richer parts of the city where sewers did not exist or malfunctioned, and in the badly ventilated homes of the well-to-do. In the nineteenth and early twentieth centuries, European travelers to the Orient and Africa added significantly to the literature on pungent and noxious odors. "No account of India," writes Porteous, "from Kipling to the recent popular novels of M. M. Kaye and the accounts of Geoffrey Moorhouse, fails to invoke the peculiar smell of that subcontinent, half-corrupt, half-aromatic, a mixture of dung, sweat, heat, dust, rotting vegetation and spices."[56]

City people have often noticed and recorded the richness of smells in the countryside: the strong whiffs of the organic manure that is amply spread over the soil on a Chinese peasant farm, the odors of manure and of farm animals in rural Europe. Although city people gladly bring home pictures of the countryside's views, they have no desire to bring home traces of its organic odors. Yet they may well appreciate these same odors in the places where they are perceived to belong. To a degree, this attitude applies also to the strong, distinctive regional odors to be found in Third World countries. Distinctive odors that assail the nostrils make a place seem real in a way that visual images alone do not. Moreover, complex odors of heat, sweat, and spice yield an agreeable impression of vitality.

It is easy to adapt to odor. We open the door of a bakery and are overcome by the fragrance of baked goods, yet a few minutes later we no longer notice it. Similarly, bad odors disappear after we have been exposed to them for a short time.[57] Odors are more likely to be noticed when they are periodic. Perhaps for this reason, the onset of spring, at the end of a long and relatively odorless winter, has inspired many eloquent literary evocations of fragrance. Tolstoy describes the excitement he felt as a child after a spring thunderstorm: "On all sides crested skylarks circle with glad songs and swoop swiftly down. . . . The delicious scent of the wood . . . , the odour of the birches, of the violets, the rotting leaves, the mushrooms, and the wild cherry is so enthralling that I cannot stay in the brichka."[58]

The traditional countryside of villages and farms also has a periodic odor that most travelers have commented on with pleasure: the fragrance of burning wood as local farmers and herders settle down in the evening to prepare their meals. This smell has been noted in the byways of Europe, Africa, and south Asia, its distinctive scent varying with the type of wood burnt. One traveler in Spain describes the particularly fine fragrance he encountered in Andalusia:

> The goats and the cattle were being driven home, the voices of the men and women calling to one another lay like long streamers on the air. . . . But what was that sweet aromatic smell? Looking round me, I saw that every one of those flat gray roofs had a small chimney projecting from it and that from each of these chimneys there issued a plume of blue smoke which, uniting with other plumes, hung in a faint haze over the village. The women were cooking their suppers and for fuel they used bushes of rosemary, thyme, and lavender which were brought in on donkeys' backs, from the hills close by.[59]

Another seasonal odor, much praised by writers and nature lovers of the eighteenth century, is the odor of newmown hay. It is not a flowery or obvious fragrance, and yet it has an immediate appeal, based on or strengthened by association with the idea of nature, of health and the outdoors, of freshness and youth. Newmown hay retains its attraction to city dwellers today—an attraction touched by the wistful awareness that, in an increasingly urbanized world, waking up to the fragrance of newmown hay is becoming extremely rare.[60]

FRAGRANCES IN THE CITY

Cities have seldom smelled pleasant. Urban boosters rarely draw on the testimony of the nose. Good odor cannot really be expected in places where humans congregate in large numbers. In the past, the odors that repelled were organic in origin. As hygienic measures brought them under control, new offensive odors—automobile and industrial fumes—assaulted the nostrils. A city may retain a certain pleasing fragrance as the result of traditional practices that have no place in a modern city. Thus certain quarters of Rome are still scented by outdoor woodfires despite municipal prohibitions. Or a city gains a distinctive aroma thanks to a particular industry. Baltimore, for example, is favored by the pungent odor from its spice factories. Certain lumber towns are enveloped in the clean smell of sawdust. In a sophisticated metropolis, a shaded street, lined by a variety of shops, can offer a real treat to the connoisseur of fragrance. Strolling along, we may pause by a fruit-and-vegetable stand to inhale the tangy aroma from the crates of oranges and lemons, mixed with the earthier odors of cabbages and potatoes; next to it, a tray of secondhand books exudes a papery and faintly musty scent of inexplicable charm to the bibliophile; a coffeeshop spills samples of its aroma into the open air of the sidewalk, as does a shoe shop when its door swings open to admit customers; and from the well-dressed and coiffed women on an upscale street we catch whiffs of expensive perfume. Vendors contribute significantly to a great city's olfactory ambience. New York's air has recently been sweetened by the aroma of honey-roasted peanuts. "The vendors, with their deep-bellied copper pans and hidden heat sources, churn together a mixture of peanuts and sugary sauce that produces a wonderful smell. Not since the yam sellers of the first half of the century has a group of vendors so enriched the air we breathe."[61]

One measure of the difference in importance between sight and smell is the extent to which both are used in the planned environment. For those designing the ideal city, visual appeal is preeminent. Even much of the tactile-kinesthetic value of architectural design is derived through the eye: through sight rather than touch and imitative bodily posture we experience the smoothness or roughness of a wall, the repose or thrust of a roof. Nevertheless, odor is a component of environment, and major high cultures since antiquity have tried to control it and to enhance the aromatic quality of place through the burning of incense. The original purpose of burning incense was to dispel baneful spirits and, more commonly, to offer acceptable prayers to the gods.[62] But the fact remains that fragrance in a censer diffuses through a building so that, in time, even when no incense is burning, a subtle aroma pervades it and its furnishings.

Buildings can be designed to be aromatic. The Babylonians were among the pioneers of aromatic architecture. Sargon II (722–705 B.C.) used cedarwood for his palace at Khorsabad, as did Sennacherib (705–681 B.C.), who planned it so that an agreeable odor emerged when doors opened or closed. Solomon's temple made use of the cedars of Lebanon. The temple in which Christ taught, rebuilt in 20 B.C., was fragrant, thanks again to Lebanon's cedars. Some Indian temples were known as "houses of fragrance." Screens of vetiver rootlets were placed over the openings of a verandah and dampened with water. "As the breeze entered the interior of the palace or temple, it would be perfume-laden and cooled. Similar mats, *khus tattis*, were woven to incorporate the scent of vetiver in the room."[63] Important entryways might be made of sandalwood, which had the added advantage of resistance to termites. In China, the camphor tree, whose wood is both fragrant and termite resistant, became a popular building material from the third century onward. Architects used it to make elegantly sculptured panelings and lattices in palaces and temples, and also for the rafters of arcades, under which the literati might stroll, protected from the weather, to enjoy the view and the wood's delicate perfume. The emperor K'ang-hsi constructed a vast complex of stately apartments in 1703 in southern Manchuria. In one of these, the wood *Machilus nanmu* was used for the beams and paneling, which were deliberately left unvarnished and unpainted so that the wood's cedarlike scent might emerge.[64]

In all parts of the world, potted flowering plants have long been brought into buildings to impart a desirable aroma to the rooms. More unusual and ambitious is to surround a building with fragrant plantings so that their odor may penetrate into the interior with the help of either a natural or an artificial breeze. The Palace of Coolness, built in H'ang-chou, China, in the thirteenth century provides an outstanding example of this practice. It was made of ivory-white Japanese pinewood. "In front of it were several ancient pine trees. An artificial waterfall cascaded into a lake covered with pink and white water-lilies. In the vast courtyard surrounding the palace were hundreds of urns containing jasmine, orchids, pink-flowering banana, flowering cinnamon and other rare and exotic flowering shrubs. They were fanned by a windmill so that their fragrance should penetrate within the great hall of the palace."[65]

AROMATIC GARDENS

When we think of an aromatic place, we immediately think of a garden. The archetype of the garden is Eden or paradise. Images of paradise differ according to culture and historical period. Not surprisingly, given the abundance of aromatic plants in the Indian subcontinent, Hindu and Buddhist paradise gardens are more redolent than those of the Islamic or Christian tradition. A description of an Indian heaven may begin with visual and auditory rewards and end with fragrance: ". . . This celestial abode is adorned with lotus lakes, and meandering rivers full of the five kinds of lotus whose golden petals, as they fade, fill all the air with sweet odours."[66] However, the many images of paradise in the Koran emphasize touch and taste—the coolness of water or air, the soft couches, the abundance of delicious fruits, the rivers of milk and wine—but seldom mention aroma.[67]

Christian images of paradise have also varied widely over time, but in most sight has been primary. Jesus' heaven is so God-centered as to leave little room for mere sensual delight. The book of Revelation emphasizes sight and sound—rainbow-colored light and winged spirits singing a never-ending "Holy, holy, holy is the Lord God Almighty." To Augustine, eternal bliss consists in "seeing God." Dante's paradise is a visual world of circling light and color. In the early Middle Ages, however, odor and nature became more salient in paradisiacal images, perhaps because monasteries, located in the odoriferous countryside, enjoyed exceptional power and influence then. According to a ninth-century monk and poet, in paradise "lilies and roses always bloom. . . . Their fragrance never ceases to breathe eternal bliss to the soul." A work called *Elucidation*, compiled around 1100 and copied often throughout the medieval period, depicts the new earth after the Last Judgment as a fragrant and pleasant garden. It will be irrigated with the blood of the saints and "decorated eternally with sweet-smelling flowers, lilies, roses, and violets that will never fade."[68]

No survey of aromatic ideal gardens would be complete without mention of Milton's Eden in *Paradise Lost*, which he began in earnest in 1658 after becoming totally blind. Arriving in Eden, the archangel Raphael

> *. . . now is come*
> *Into the blissful field, through Groves of Myrre,*
> *And flouring Odours, Cassia, Nard, and Balme;*
> *A Wilderness of sweets; for Nature here*
> *Wantoned as in her prime, and plaid at will*
> *Her Virgin Fancies, pouring forth more sweet,*
> *Wilde above rule or art; enormous bliss.* (V. 291–297)

Real gardens cater primarily to the eye. The visual emphasis is especially salient in the largest and most resplendent gardens of the Western world, such as those at Versailles, which have been designed as symbols of prestige, stages for pageantry, settings for amateur performances and spectacles. In China, too, since the eighteenth century connoisseurs have pretended to see gardens as landscape paintings—visually desirable objects—that one can miraculously enter. However, under the influence of Taoism and Buddhism, the garden was and is more importantly a place of retirement, repose, and contemplation. There eyes could turn inward as much as outward; they could be figuratively and literally half-closed. With the visual sense thus withdrawn from its position of dominance, the other senses—including smell—could be heightened. The plants in a Chinese garden are valued not only for their shape, color, and symbolic resonance, but also for their distinctive odors; thus pavilions might be located with sources of fragrance in mind—perhaps downwind from a lotus pond or in a pine grove.[69]

Enclosed gardens of modest size are common in many parts of the world besides China: for example, in countries that have come under Islamic influence. But even in the West, great gardens that appeal first to the eye may also boast small enclosures for retirement and repose. In such an enclosed space, in a relaxed mood, seekers after peace are able to enjoy subtleties of sound and odor—the gentle twang of zither and the chirping of birds, the scent of herbs and flowering plants. In eighteenth-century Europe, the sequestered parts of gardens were designed with mounting attention to fragrance. In "serene regions" bordered by plants with odoriferous flowers or foliage, sweet, balsamic aromas would promote a tranquil state in which people gave themselves over to soothing sensations and reflection rather than to thrilling views and purposeful thought.

ODOR BIASES AND PREDILECTIONS

Attitudes to odor are complex, ambiguous, and emotional. In the West, we have seen how a negative attitude developed in relation to the belief that nature's odors could cause harm. The human body is a part of nature. Tolerance for its odors, which can be pungent after hard physical work, varies from culture to culture. Many nonliterate peoples bathed frequently, as did the ancient Egyptians, Greeks, and Romans. By contrast, during the long period from the end of antiquity to modern times, even the elite of European society showed little interest in personal hygiene. They sought to counteract the effects by drenching themselves in perfume. From the late eighteenth century onward, with the development of medical science and engineering technology, members of the middle class became increasingly sensitive to odors of organic decay and physiological functioning, and tried hard to eradicate them from their houses and streets, as well as from their own bodies. One result is an ambivalence about odor itself, even odor considered pleasant.

Today, especially in Anglo-American society, people prefer odors that are subtle and faint. Some regard even the smells of a flower shop, greenhouse, bakery, leather goods store, and ethnic eating places a little too strong. At the other extreme, a completely odorless place, such as a camera shop, produces a somewhat sterile, cool, and distancing effect. In fact most offices, public spaces, and stores that cater to the middle class aspire to near odorlessness. The exceptions—grocery stores, old-fashioned general stores, bookshops that have a mixture of old and new books, the cosmetics section of a department store—have aromas that are pleasing not only because they are faint and complex but also because they are distinctive. Merchants now realize that odors can enhance the attractiveness of merchandise made of natural materials, making it seem more authentic, since almost all natural objects, including rock, have odor.[70]

As for the distinctive fragrances of natural landscapes, olfactory geography remains strikingly underdeveloped. People grow accustomed to the sights, sounds, and smells of their own homeplaces. Smells, in particular, tend to be taken for granted. When people move out of their familiar milieu, they may be rudely shocked by novelty and intensely miss the sensory worlds they have left behind. To hunter-gatherers of the tropical rain forest, the steppe seems barren and hostile not only because of its vast horizon but also because it is far less odoriferous; Hawaiians reportedly miss the pungency of their islands when they visit the mainland's less aromatic Midwest, and Midwesterners may feel overcome by tropical scents when they arrive in Hawaii.

Scents capture the aesthetic-emotional quality of place. In middle-latitude deciduous forests, it may be the smell of spring, understood as the smell of life; in the boreal forests, the smell of pine needles and fir cones; in certain parts of the Great Plains, the fragrance of sage; in New Mexico, piñon and juniper basking in the hot sun; in Australia, eucalyptus; on sea coasts, cool ozone-tinted air with a whiff of seaweed; in hot deserts, the slightly acrid odor of intensely heated soil and rock; and so on. People who speak of preserving natural landscapes usually have only their visual character in mind. Yet not only a silhouette but an aroma is destroyed when an old deciduous forest is chopped down and replaced by fast-growing conifers. An olfactory geography does not come naturally to us. For this reason, when we have learned to appreciate nature's extraordinary range of aromas and are ready to write an olfactory geography, we will have demonstrated a higher level of aesthetic sophistication.

Voices, Sounds, and Heavenly Music

*M*ost people consider loss of sight a greater catastrophe than loss of hearing. After all, we have only to close our eyes, and we are instantly plunged into darkness; whereas if we plug our ears, the resultant silence may seem peaceful, as a city is peaceful, its noises muffled, after a heavy fall of snow. Moreover, absence of sound can enhance visual acuity: we focus better and the world seems more sharply defined without the distractions of sound.

But soundlessness begets a sense of deadness. Life involves movement, and when living things move, they generally make some kind of sound. Soundless movement can seem uncanny, ghostlike, and vaguely threatening. In my soundproof high-rise apartment I stand by my picture window and look out to the cityscape below. I see buildings and streets with people and cars moving on them. I can see that the city is alive, but I know it—I feel it with my whole being—when I throw the window open and street noises rush in.

The Human Voice and Other Everyday Sounds

Of the many varied sounds in our environment, by far the most important, and therefore the one to which we are most sensitive, is the human voice. We become human by listening to others speak and learning to speak; we become participating members of a community primarily through speech, with its infinitely nuanced messages of love and hate, admonition and commendation. Those deprived of hearing are isolated from this reassuring cocoon of voices.

Human voices are mixed in with other sounds that are almost as familiar and reassuring. We cannot be totally alone and retain a sense of who we are. Other presences are needed—as John Updike puts it, "soft night noises—a mother speaking downstairs, a grandfather mumbling in response, cars swishing past. . . . We need all the little clicks and sighs of a sustaining otherness. We need the gods."[1] Urbanites who tire of the bustle and raucous noise of the city seek the peace of nature—peace rather than silence, for in nature we hear the muffled roar of falling water, the rustle of leaves, the chatter of birds. Absolute silence is intolerable. In the midst of nature, much as we take delight in the call of the loon and other anticipated sounds, we are supported even more, at a deeper level and as a cushioning backdrop, by the innumerable little clicks and sighs that envelop us.

Biological Rhythms

Humans may be primarily visual, but hearing enjoys temporal priority. Infants in the womb can hear their mother's breathing and heartbeat, digestive processes, and muscle and joint movements. They also register messages from the external world, such as loud conversation; even before birth they may become accustomed to their parents' voices.[2] Thus infants acquire their earliest awareness of the external world through organic and human sound rather than through sight. The steady sounds of organic life, especially the heartbeat, provide comfort. Even a left-handed mother tends to hold her infant in her left arm, apparently from the instinctual knowledge that the baby appreciates the rhythmic throb of the heart. Adults retain this preference. "Wherever you find insecurity," Desmond Morris asserts, "you are liable to find the comforting heartbeat rhythm in one kind of disguise or another. Rocking back and forth. . . . It is no accident that most folk music and dancing has a syncopated rhythm. . . . It is no accident that teenage music has been called 'rock music.'"[3] According to musicologist Murray Schafer, the fundamental drumbeat of Australian aborigines hovers around the heartbeat; so does Beethoven's Ode to Joy. The heartbeat provides a measure for the human understanding of fast and slow: as military music pushes up the tempo, people move faster, with a greater sense of purpose. Breathing provides another such measure. According to Schafer, one reason why people feel good at the seaside is that the breakers' rhythms, though never regular, last around eight seconds each—an interval that corresponds to that of relaxed breathing.[4]

Sound, Emotion, Aesthetics

Sound can arouse human emotion to a more intense level than can sight alone. "Screaming" headlines in the morning newspaper catch our attention but have no grip on our heart. Pictures of disaster may elicit more of a response. But we will be thoroughly engaged by the sound of an ambulance siren or by cries of pain, rage, or despair.

Sound and smell are cues to the environment, informing animals about sources of food, sexual opportunity, and danger. Humans depend less on hearing and smell than do nonprimates, who most actively face the pleasures and dangers of life at dusk, dawn, or even during the night. Sound does still serve as an important source of warning: thunder signifies potential danger; a camper in the wilderness grows sensitive to the sound of twigs breaking as a possible indication of an approaching bear; at home, one asks in the middle of the night, "What is that noise in the kitchen?" However, our greatest aural sensitivity is to the volume and quality of other people's voices. The meanings of an infant's cry, a raised voice, and shouts of anger are unmistakable, but beyond these, people are remarkably good at discerning finely shaded degrees of hostility and friendliness from the tone of the voice, irrespective of the words' literal meaning.

Are the deeply felt emotions evoked by sound aesthetic? Certainly, sound is aesthetic in the sense that movement itself seems "dead" without it. On the other hand, distancing is generally more difficult in an auditory world than in a visual one: sounds tend to wrap around us. Even so, we rarely lose ourselves in sound so completely that we cannot step aside briefly to listen. In the midst of a rich aural environment, we have no trouble savoring its distinctive sounds, in isolation or in symphony—the cooing of pigeons, the lapping of waves on the seashore against the distant sighing of wind, the sweet lilt in the voice of someone we love, the pure note of the violin soaring above the orchestral groundswell.

Ominous Silence

Perhaps because they almost always reach our consciousness as discrete events, sounds seem to exist for us against a background awareness of nonexistence—that is, of silence. Indeed, part of the attraction of sound, such as the lone bird call in the prairie, is the silence that immediately precedes and follows it. In our frantic noise-filled modern world, silence is considered highly desirable—an aesthetic, life-restoring experience in its own right. Silence, however, also signifies death. Silence is of the tomb. In Greek myth, Orpheus descends into the Hades of darkness and silence, which he temporarily conquers with his music. From Greek antiquity to the threshold of the modern period, Europeans entertained the pleasing and reassuring idea that the heavenly spheres were filled with music. As belief in the idea waned, the heavens seemed increasingly unresponsive, cold, and silent. The mathematical genius and mystic Blaise Pascal was frightened by "the eternal silence of these infinite spaces." For the Hebrews, "In the beginning was the Word." Without the Word of God "was not any thing made that was made." In the Word "was life, and the life was the light of men." God brought a world into being by utterance, which broke the void, darkness, and silence of primeval chaos. We still think of speech as creative and healing. Silence is darkness, language is light.

> *Der Mond ist aufgegangen,*
> *die goldenen Sternlein prangen*
> *am Himmel hell und klar;*
> *der Wald steht schwarz und schweiget,*
> *und aus den Wiesen steiget*
> *der Weisse Nebel wunderbar.*

> *(The moon has risen,*
> *The golden starlets shine*
> *In heaven bright and clear;*
> *The forest stands dark and silent,*
> *And from the meadow ascends*
> *An enchanting white mist.)*

In this poem by Matthias Claudius (1740–1815) "the demonic silence of night is overcome by the brightness of language," writes Max Picard. "Moon and stars, forest, meadows and mist all find and meet each other in the clear light of the word. . . . Through the word the silence ceases to be in demonic isolation and becomes the friendly sister of the word."[5] Silence is bafflement, awaiting the clarification of speech. Silence is the indifference of powerful people. It is also the indifference of nature.

As Pascal uncannily foresaw, the silence of outer space would be oppressive, frightening. Imagine the terror of an astronaut, cut off accidentally and irremediably from his spacecraft, drifting into the void and silence of space—to death, utterly alone. Here on earth, too, the silence of nature can overawe or induce a sense of the uncanny. When Europeans first encountered the North American prairie, they found it forbiddingly strange, and one component of that strangeness was its silence. "It is truly remarkable," wrote the botanist Thomas Nuttall in 1819, "how greatly the sound of objects becomes absorbed in these extensive woodless plains. No echo answers the voice, and its tones die away in boundless and enfeebled undulations."[6] New World forests, too, differ from those of settled Europe in the volume and quality of their sounds—and in the brooding intensity of their silence.

A common source of sound in nature is the impact of wind on vegetation. Where there is little wind and little or no vegetation, silence reigns. The great hot deserts of the world often fall silent. Under the relentless sun and a sky bleached white by heat, when nothing stirs except an occasional dust devil, one may hear—in a condition of exhaustion mixed with panic—only the thump of one's own heart and a hum in the ear. The great ice caps can be even more awesomely still, uncannily quiet. Richard Byrd, camping alone on the ice plateau of Antarctica at latitude 80°08' south, wrote in his diary on May 1, 1934: "The winds scarcely blew. And a soundlessness fell over the Barrier. I have never known such utter quiet. Sometimes it lulled and hypnotized. . . . At other times it struck into the consciousness as peremptorily as a sudden noise. It made me think of the fatal emptiness that comes when an airplane engine cuts out abruptly in flight."[7]

Noisy Nature

Nature can, of course, also be full of sound and fury: think of the peal of thunder, the screech and roar of a tropical storm, the bulletlike bombardment of hailstones on the roof. A needle-leaf northern forest can be eerily quiet, but all this changes when the wind starts to blow: the wind, gentle at first, produces a low breathy whistle, but as it gains strength the billions of needles begin "to twist and turn in turbine motion to generate a seething roar."[8] A geologist friend of mine, tired of the noise of New York whence he came, and dissatisfied even with the relative quiet of Albuquerque, New Mexico, where he was studying, decided one summer to do fieldwork in the high mountains of Peru. When he returned, I asked him what it was like to be utterly alone among the ice-covered peaks. He said he had had difficulty sleeping at night: the periodic roar of the avalanches made him feel that he had camped in Grand Central Station.

Even Antarctica and its waters are by no means always quiet, despite the absence of strong wind. Wherever the ice is under pressure, it creaks and groans, or detonates; under more complex conditions it can make a weird assortment of noises. One member of Shackleton's expedition to Antarctica described the Wendell Sea on March 11, 1915:

> The area of dangerous pressure . . . does not seem to extend more than a quarter of a mile from the berg. Here there are cracks and constant slight movement, which . . . makes all sorts of quaint noises. We heard tapping as from a hammer, grunts, groans and squeaks, electric trams running, birds singing, kettles boiling noisily, and an occasional swish as a large piece of ice, released from pressure, suddenly jumped or turned over.[9]

Nature also produces steadier sounds: the constant roar of a large waterfall; the thud-and-swash of waves on an exposed coast; and, in the tropics, the sound of heavy rainfall beating for days at a time on leaves and roofs. And then there is the wind. In the Bahamas, the trade winds are the islanders' constant companion. One islander describes them as the third force, the other two being the firm earth and the surging sea: "This third force seemed to come out of the emptiness of outer space; the wail of its pressure dominated the whole of existence. It did not rise and fall in rhythmical sequence like the surf but maintained a constant tenor, a deep organ-like lamentation that swept on and on in constant reiteration, never ceasing."[10]

Animal cries add to nature's pandemonium. In the urbanized modern world, they tend to be kindly received because they suggest the countryside and wilderness. Of course, such tolerance can reach its limits with the annoyances of the barking of the neighbor's dog, the buzz of flies, the whine of mosquitoes. But not so long ago, even in well-settled western Europe, animal cries could cause deep anxiety—especially wolves, driven out of the forest by hunger during long, harsh winters and converging on villages and towns. In our own century, the peasant farmers of Uganda dreaded the sound of wild trumpeting elephants crashing through the forest, destroying the fields that lay in their path, threatening villagers with injury and death.

In general, the richer a biotic community, the noisier it is: at one extreme, the silence of the great deserts and ice caps, and at the other extreme, the raucousness of tropical forests. The animal noise level of middle- and high-latitude forests lies somewhere in between. And there are differences even here: in North America, the northern conifers are quieter than the eastern deciduous hardwoods. Certain parts of the world are exceptionally strepitous by virtue of the presence of *one* animal species. In Australia, for example, the unceasing buzz of the cicadas, the loudest of insects, adds to the oppressiveness of summer; at night, when they quiet down, Australians find some relief in the less insistent, gentler warbling of crickets.[11] The tropical rain forest, its floor illuminated by flickering patches of sunlight, at times is as quiet and cool as the vaulted interior of a cathedral; at others, it may be filled with raucous din, raised to a volume unimaginable to wilderness lovers of higher latitudes. Here is an account of what can be heard in the Ituri forest of Zaire.

> The darkness quaked with frog and insect din. A troop of black-and-white colobus monkeys . . . called to each other. They sounded like several motorcycles being revved. Every minute or so, a bloodcurdling scream, as if from a woman about to be murdered, would sound in a nearby treetop. Its perpetrator was the tree hyrax—a small, edible gray mammal with a white dorsal tuft, whose family is most closely related to elephants.[12]

Nature's Pleasing Sounds

Nature produces not only frightening, oppressive, or irritating sounds but also reassuring, invigorating, or delightful ones. Although everyone no doubt has his or her favorites, the literary evidence suggests that there is also substantial agreement on the sounds that are reassuring—that lull us like lullabies, but without their sentimentality: the sighing of gentle rain, the rustling of autumn leaves, waves lapping on the lakeshore, gurgling water, pigeons cooing in a misty morning.

Sounds that emerge unexpectedly out of silence and quickly return to it catch our attention. Such sounds may at first cause mild alarm, for we are adapted to interpret them as warnings. On the other hand, they can also offer pure pleasure, even mystery and exhilaration. The Canadian musicologist Murray Schafer offers several examples. On a calm day high in the Swiss Alps, across the intervening stillness of the valleys, one may hear unexpectedly a whistling sound, created by wind rushing down a glacier miles away. To hear the wind and not feel it creates a sense of mystery; when the sound fades, the region's stillness seems enhanced.[13] And in the vastness of the Russian steppes, the isolated song of a single species sometimes blends with the sound of sleigh bells in winter. Maxim Gorky asks, "What could be more pleasant than to sit alone at the edge of a snowy field and listen to the chirping of the birds in the crystal silence of a winter's day, while somewhere far away in the distance sounded the bells of a passing troika—that melancholy lark of the Russian winter?"[14] Boris Pasternak, too, writes of the beauty of single notes in open space: "Everything might be dead; only above in the heavenly depth a lark is trilling and from the airy heights the silvery notes drop down upon adoring earth, and from time to time the cry of a gull or the ringing note of a quail sounds in the steppe."[15]

The countryside is alive with sounds. Particularly joyful are those heard at the coming of spring. Tolstoy's works are full of vivid evocations of nature, turning over majestically as one season follows another.

> The old grass turned green again and the young grass thrust out its needle-sharp blades. . . . In the gold-besprinkled willows the honey bee, which had only just emerged from the hive, flew about humming. Invisible larks broke into song over the velvet of the young, sprouting corn and the ice-covered stubble; peewits began to cry over the marshes and the low reaches of the rivers and streams . . . and cranes and wild geese flew high across the sky, uttering their loud, spring cries. The cattle, their winter coats

only partly shed and bald in patches, began to low in the pastures; bandy-legged lambs frisked round their bleating mothers, who were losing their fleece. . . . Real spring had come.[16]

In the intervals of stillness, the subtler sounds of life rise to the foreground. "How do you like that! One can see and hear the grass grow, Levin thought to himself, noticing a wet, slate-colored aspen leaf moving near a blade of young grass. He stood listening and looking down now on the wet, mossy ground."[17] According to Francis Kilvert, crops have different "voices," depending on whether it is day or night. In his diary for July 16, 1873, he wrote: "As I walked along the field path I stopped to listen to the rustle and solemn whisper of the wheat, so different to its voice by day. The corn seemed to be praising God and whispering its evening prayer."[18]

Human Cacophony

Before people began to live in large cities, noise was nature's prerogative. People might shout and beat drums, but the level of noise they made seldom matched that of thunder or violent wind. The gods of nature were jealous of their exclusive right to make noise, as the following story about the origin of flood, taken from the Sumerian epic of *Gilgamesh*, illustrates.

> In those days the world teemed, the people multiplied, the world bellowed like a wild bull, and the great god was aroused by the clamour. Enlil heard the clamour and said to the gods in council, "The uproar of mankind is intolerable and sleep is no longer possible by reason of the babel." So the gods in their hearts were moved to let loose the deluge.[19]

Enlil was the Sumerian god of wind and storm; his roar manifested his power. Humans, by bellowing "like a wild bull," not only kept Enlil awake but also challenged his divine prerogative to overwhelm with noise. Jehovah, God of the Hebrews, came from the same cultural area as the Mesopotamian nature gods. He spoke out of the whirlwind; he, too, saw fit to wrap himself in the power and majesty of motion and sound.

In premodern battles, noise was an important part of the strategy to unnerve the enemy: shouting, drum beating, saber rattling, even the firing of cannons might not necessarily kill, but they did project an impression of irresistible force. Significantly, from the fifteenth to the eighteenth century cannons were sometimes decorated with animal motifs: the cannon roared like a lion—it simulated nature's own violence. Similarly, immense noise in large cities signaled simply tumultuous activity. Yet authorities viewed the clamor with suspicion, for it portended anarchy and chaos; it expressed raw energy, like that of nature's storms, which could veer out of control.

To city people since ancient times, noise has been a distressing but inevitable aspect of the environment, to be complained about and adapted to. In imperial Rome, for example, although citizens could find oases of beauty and calm in the city's more than forty parks and gardens, more often they had to contend with the breathless jostle and infernal din of its narrow streets. The presence of numerous traders contributed to the intense animation and cacophony with the cadence of their tools, the rush and hustle in their toil, and their loud swearing. Night brought no peace, because it was then that wheeled carts could legally enter the city. Juvenal asserted that nighttime traffic condemned Romans to everlasting insomnia.[20]

Bustling activity and noise enlivened and afflicted all crowded cities in premodern times, although the kinds of noise differed with locality and with changes in technology. Thus, in medieval European cities, church bells tolled almost incessantly and could be annoying even to people habituated to them. By the eighteenth century, the sound of bells gave way to the deafening roar of wheeled traffic. Wheeled carriages began to move into the urban core in the sixteenth century. Streets were broadened to accommodate them. More and more vehicles cluttered up the public arteries, creating a continuous racket. In colonial America, too, cities became noticeably noisier. Travelers remarked on the sharp contrast between the quiet of the countryside and the hubbub of the towns. When botanist James Young entered Philadelphia one July day in 1763, he found himself "tangled amongst Waggons, Drays, Market Folks and Dust." A medical student living on Second Street wrote home and decried "the thundering of Coaches, Chariots, Chaises, Waggons, Drays, and the whole Fraternity of Noise [which] almost continuously assail our ears."[21] In 1771, London had a thousand hackney coaches. These were at first heavy vehicles with perforated iron shutters. Their wheels grinding on cobbled streets produced an abhorrent din.[22] Shopkeepers complained bitterly but to no avail. A century later, traffic noise was even worse; not only had the number of vehicles increased, but their heavy wheels as yet unshod by rubber still bore down on streets paved with stone blocks. "In the middle of Regent's Park or Hyde Park," a Victorian gentleman recalled, "one heard the roar of traffic all round in a ring of tremendous sound, and in any shop in Oxford Street, if the door was opened no one could make himself heard till it was shut again."[23]

Except for those that still have traditional marketplaces and bazaars, cities today are comparatively quiet. True, certain corridors are assaulted by the roar of airplanes passing overhead and the noise of vehicular traffic on the ground. But car horns no longer sound incessantly in North American cities, as they did in the 1920s and 1930s; drivers now press the horn only in an emergency. Indeed, on the streets of many North American cities the noises of life seem to disappear not only late in the evening but also in daylight, on weekends, and on public holidays. Then, even a metropolis the size of Dallas turns into an eerily silent sculptural garden.

The City's Pleasing Sounds

Dynamism imbues the hum and roar of urban life. Under certain conditions, even cacophony is enjoyable: during the rush hour, to a pedestrian if not to the driver immobilized by traffic, the crash of dissonant sounds (with an occasional consonant beep from car horns) is not wholly without aesthetic appeal.[24] Indeed, some cities seem to have their own special sonic ambience. The Danish architect Steen Rasmussen writes: "From my own childhood I remember the barrel-vaulted passage leading to Copenhagen's old citadel. When the soldiers marched through with fife and drums the effect was terrific. A wagon rumbling through sounded like thunder. Even a small boy could fill it with a tremendous and fascinating din—when the sentry was out of sight."[25] A Copenhagen child will find that certain walls in the city offer special satisfaction for ball bouncing: the ball, as it hits a surface of a certain hardness and texture, makes a sweet sound—a "boing!" that stimulates the child's senses and at the same time imparts to him a feeling of crisp competence.

Agreeable urban sounds are unobtrusive. We seldom pause to consider their role in the daily theater of life. Consider the following sequence of events, which unfolds every day. Very early in the morning, a public square is still in shadow, starkly empty, and quiet. As the sun rises and progressively illuminates the square, we hear a succession of sounds—first the swish of water expelled from the tank of a sanitary truck, then the tapping of footsteps separated by diminishing intervals of silence; a bell rings the hour, voices rise against the muffled sound of motorized traffic in neighboring streets, and life begins in earnest.

In a study of the sonic environment of Boston, Michael Southworth notes that all his subjects liked quiet but resonating places such as Beacon Hill, India Wharf, or the alleys, and preferred constantly varying soft personal sounds—footfalls, fragments of conversation, whistling, or shuffling—and clear, novel, or informative sounds. He blindfolded some subjects, plugged the ears of a few others, and left the remainder unhampered. He found that the three groups differed significantly in place preference. Both auditory and visual-auditory subjects liked Washington Street, Beacon Hill, and India Wharf, all bustling places that project sonic-visual variety and distinction. Visual subjects preferred Quaker Lane and Boston Common, which looked nice but had low sonic appeal. They did not like India Wharf and made such negative comments as: "The water looks quite dull, the sad emptiness with nothing in the foreground to look at is very disturbing . . . and there is a sight-seeing boat, [but] I can't imagine why anyone would want to go out and sightsee here." With their ears plugged, they could not enjoy the sonic delights of the place: the creaking of boats, the drone of distant planes, the lapping of water, the chime of bells, the mourn of ship horns, and the cry of sea gulls.[26]

San Francisco is justifiably proud of its visual beauty: it has a dramatic geographic setting, the Golden Gate Bridge, and buildings of elegance and distinction. But it is also proud of its sonic environment. A Christmas record produced commercially in the early 1970s welcomes its listeners with the words, "Merry Christmas from San Francisco, the magic city of soaring hills and bridges that leap across the Bay . . . a city of many voices that you can hear nowhere else." Side one of the record contains the chimes of Saints Peter and Paul Church in North Beach, a streetcorner Santa Claus, a Salvation Army band, Christmas shopping traffic on Market Street, and a ride on a Powell Street cable car up Nob Hill to the tune of "Jingle Bells." Side two contains a holiday celebration in Chinatown and sounds of the bay itself: boat whistles, water lapping against a pier, the pounding surf, wind singing through the cables of Golden Gate Bridge, sea gulls and sea lions, and the many voices of the foghorns in their Christmas carol.[27]

San Francisco is exceptional in possessing a sonic personality that is distinctive and aesthetically pleasing. Few cities in the United States, or even the world, match it—certainly not Los Angeles, its rival metropolis in the south. The natural environment notably contributes to a city's sonic character: one cannot imagine San Francisco without the bay—and that includes the bay's sounds. Most cities offer few auditory rewards, and these are as likely to be natural as manmade. Are there workaday artifactual sounds with enough appeal to make people pause and listen, as they do when they catch the warbling of a lark? The bell is a possible candidate. Many people like its almost-human sound. In medieval times, church bells were treated as almost human; they were consecrated and given personal names. The bell was the voice of Mother Church, concernedly ordering the lives of her children. Indeed, when merchants in fourteenth-century France built belfries and put in bells to regulate the schedule of their workers, the workers rebelled successfully with the support of the Church, which did not welcome a rival voice.[28]

Over a vast expanse of still and calm space, such as the steppe or the inland sea, the sudden cry of an animal (wolf, sea gull, lark) can evoke deep, almost primordial emotion in the human being. Curiously, this is also true of certain sounds that come out of manufactured objects. Think of the shiphorn or foghorn or the train whistle. Mixed in with other sounds, they may have no special significance—they are just noise. But alone, they can strike us as intensely human. They are like a cry, a mournful and soulful expression of loss or yearning that deepens the circumambient silence. Such sounds bring us to the verge of the mystery of music—especially the most developed form of music—a dynamic patterning of sounds that we listen to, serene or in rapture, for itself.

Music

Nature offers a vast range of sounds, some of which people have learned to appreciate for their beauty and power, and not simply because they indicate a possible source of danger or food. Nevertheless, the sounds most enthralling to humans are those they have made themselves. In this bias in favor of the humanly created, hearing differs from touch, smell, and sight. The tactile pleasure of running our hand over the polished surface of a mahogany table is no greater than that of feeling the smoothness and warmth of a sun-baked pebble. Perfumers' exquisite scents are not manifestly superior to natural scents, on which most perfumes, in any case, depend. As for sight, we feast on great works of art, yet in its power to enchant, nature can more than hold its own: Michelangelo's David is matched by Davids who strut the streets of Florence, and the best landscape paintings in the world evoke an aesthetic response no more intense than that obtainable from nature. Taste, on the other hand, favors the artifact; although fresh fruit is delicious and raw fish is considered a delicacy in some cultures, people generally prefer cooked food. In this bias toward the artifactual, taste is like hearing. Cooking and making music are among the most widely practiced of arts. Cooking enjoys prestige in some societies—outstandingly, the Chinese. Nevertheless, the prestige of music is far greater: cooking, no matter how refined, is still a service to the body, whereas music, in its exalted forms, elevates the soul.

One way to understand the importance of music is to pose it against noise and silence. Noise, the clash of dissonant elements, signifies danger and impending chaos. Silence, too, can betoken danger before a storm or a burst of anger; it can also suggest isolation, death, indifference. But silence has positive meanings as well. It is the opposite of noise, the calm after a storm, the reconciliation after a quarrel, the pregnant interval filled with life and thought, quiet social cooperation. Considered thus, silence is compatible with music; indeed, the pauses between notes are integral to certain kinds of musical composition and account for their poignancy and serenity.[29]

MUSIC AS COMMUNAL CELEBRATION

Music plays an important role in almost all human groups. Consider a people of simple material culture, the Mbuti Pygmies, who are hunter-gatherers in the Ituri rain forest in Zaire. The Mbuti dislike noise, which they associate with nature in a violent mood and with human strife. They like silence. But even more they like music and the making of music. In common with other societies, musicmaking among the Mbuti serves a social purpose. It is a response to crisis and occurs when the Mbuti feel the need for what they call "curing," following a bad hunt, bad weather, or bouts of sickness. They think of their songs as "wakening" the forest to their needs. Singing is work—hard work; important, to them, is not whether one sings well, but whether one puts all one's energy into the song.

Singing unifies the group at a deep psychological level. Colin Turnbull, who attended a Mbuti songfest, closed his eyes to listen better, and he joined in the singing so as to feel himself more a part of the group. When he opened his eyes, he "saw that while all the others had their eyes open too, their gaze was vacant. . . . There were so many bodies sitting around, singing away, but I was the only person there, the only individual consciousness, all the other bodies were empty." Singing not only welds the human participants into a communal whole; it unites that whole to a larger whole, which includes "the central fire, the camp itself, the clearing in which the camp is built, the forest in which the clearing stands, and whatever . . . contains the forest." And it "very definitely includes whatever is implied by such equally ambivalent terms . . . as God and Spirit."[30] To the Mbuti, certain songs carry supernatural power. Their elevated standing is captured by a popular legend, known as the Bird with the Beautiful Song. The bird sings the most beautiful song that the forest has ever heard. A man, in a fit of frustration and anger, kills the bird. Immediately he himself falls down dead, "completely dead, dead for ever," as the legend puts it.[31]

Music invites participation. It is communal work, celebratory utterance and prayer; and although it may have a specific, externally directed purpose such as praising God or procuring abundant harvest, its most certain effect is internal; that is, it strengthens group feeling. Almost all premodern societies encouraged group singing. The early Church Fathers urged worshipers, even those without vocal training or talent, to participate in song. St. John Chrysostom was typical: "Even though the meaning of the words be unknown to you, teach your mouth to utter them meanwhile. For the tongue is made holy by the words when they are uttered with a ready and eager mind."[32]

In entertainment or art, spectators and listeners are separated from actors and artists. By contrast, in folk festivals and religious rites, although individuals do occasionally stand aside to listen and observe, judge and evaluate, they do not constitute a separate group, and they tend to see themselves as participants even when they are not actively involved. Thus Victor Zuckerkandl says of the Gregorian chant that it has "not an audience but a congregation. They have assembled not to listen but to worship. The chant is not sung to them but for them, on their behalf. The division into singers and listeners remains on the surface, beneath which all of them, singers and listeners alike, are one."[33] What is true of the Gregorian chant, a sophisticated form of music in a formal rite of the Church, is even more apt a description of musicmaking in folk festivities and, generally, during all those occasions when people congregate to celebrate a social or religious event. Music or chanting is a vehicle for communal rejoicing or communal pleading to the gods.

MUSIC AS AMBIENCE

In worship, music joins prayers and gestures to establish spiritual communication. Thus practiced, music is not a separate entity that demands or receives special attention. Nor is music listened to in other, more general settings in which it provides merely a pleasing ambience. In the courtyard of a Renaissance *palazzo*, one might hear twittering birds, plashing fountains, and perhaps the sound of someone plucking idly the strings of a mandolin. In a princely mansion, ladies might chat at one end of the spacious salon while a quartet of musicians played softly at another. The desire for music as background or ambience is perhaps even stronger today, in part because it is so easily accessible through the technologies of recording and in part because there is something about the modern urban environment that demands enlivening and sweetening by music. Perhaps it is the mechanical character of the noise in our time—the loud, insistent punch of pneumatic drills, the revving of motorcycle engines, the whine of police cars and ambulances—that is especially disturbing. And it may be that even more than noise, the dead quality of silence in a modern metropolis calls for a quickening of life by musical means. In the sealed spaces of high-rise office buildings and residential apartments and in enclosed shopping malls, the life-giving sounds of nature—rain spattering on the roof, wind whistling down the chimney, squirrels rustling piles of fallen leaves—are absent. Without these background noises, space feels sterile and vaguely threatening. One grows uneasy and yearns for sonic stimulation.

In our society, music fills certain rooms at home and small public enclosures such as waiting rooms, restaurants, and shops. As a rule it fills large public space only on festive occasions, when merrymakers sing and dance in the square, a band marches down the street, or an orchestra performs outdoors. But in some localities musical sounds are an element of the everyday environment. In certain old European cities and on some American college campuses, carillons in bell towers play at regular hours, unobtrusively massaging the ears of passersby and those who have the leisure to sit on a bench. In a Muslim city, the periodic chant of the muezzin calling the faithful to prayer wafts serenely over the subfusc noises of the neighborhood. On Nicollet Avenue in Minneapolis, bus shelters provide piped classical music, which is loud enough to be heard beyond the shelter's open walls, so that on a quiet automobile-free Sunday, one can hear a complete Beethoven concerto by walking down its tree-lined sidewalk. For Chekhov in 1891, Venice had a unique musical ambience:

> And the evenings! Good God in Heaven! Then you feel like dying with the strangeness of it all. You move along in your gondola. . . . All about you drift other gondolas. . . . Here is one hung about with little lanterns. In it sit bass viol, violin, guitar, mandolin and cornet players, two or three ladies, a couple of men—and you hear singing and instrumental music. They sing operatic arias. What voices! You glide on a bit farther and again come upon a boat with singers, then another; and until midnight the air is filled with a blend of tenor voices and violin music and sounds that melt one's heart.[34]

THE CENTRALITY OF MUSIC IN CHINA

Music may be important to the ordering of society, although few cultures have acknowledged this possibility explicitly. However, two ancient civilizations—the Chinese and the Greek—did develop a metaphysic of music, giving it a central role in the harmonious operations of nature and society, though in strikingly different ways.

Many of the dynastic histories of China contain a special chapter devoted to music and its functions. One scholar asserts that "perhaps no historical records of any culture that ever existed have focused so greatly on music and its place in the life of the people."[35] Under the centuries-long influence of Confucianism, the ideal scholar-official was regarded as one who knew not only *li*, propriety and the rules of conduct, but also *yueh*, music. *Li* and *yueh* were intimately paired, visibly so during ceremonials and rites. Even on purely social occasions when music was not played, it was still considered the model and soul of conduct; and when played, it was taken to be a force that made conduct not only proper and graceful but heartfelt. It has been said of Confucius that he could assess the quality of government in any place by its music. Music, in the Confucian view, could be good or bad. Bad music was of two kinds: the loud and jarring, which encouraged disorder and threatened the integrity of the state; and the pleasing but lewd, which had the power to enervate individuals and so harm society. Good music, by contrast, promoted social harmony. Indeed, it could promote harmony in the entire universe. In response to a great man's offering of music and associated rites,

> Heaven and Earth will act in happy union, and the energies of nature—now expanding, now contracting—will proceed harmoniously. The genial airs from above and the responsive action from below will overspread and nourish all things. Then plants and trees will grow luxuriantly, . . . the feathered and winged animals will be active; horns and antlers will grow; insects will come to the light and revive; birds will breed, mammals will have no abortions. . . . And all this will have to be ascribed to the power of music.[36]

THE HARMONY OF THE SPHERES

Whereas the Chinese tended to emphasize music's human origins and its role in human affairs, the Greeks saw it as a fundamental fact of the universe.[37] In the idea of the harmony of the spheres—an idea that originated with the Pythagoreans and persisted into the Renaissance and beyond—the universe itself functioned as a musical instrument. "What is this large and agreeable sound that fills my ears?" asked Cicero. Drawing on Greek thought, his answer was that it is produced

> by the onward rush and motion of the [heavenly] spheres themselves; the intervals between them, though unequal, being ex-

actly arranged in a fixed proportion, by an agreeable blending of high and low tones various harmonies are produced; for such mighty motions cannot be carried on so swiftly in silence; and Nature has provided that one extreme shall produce low tones while the other give forth high. . . . Learned men, by imitating this harmony on stringed instruments and in song, have gained for themselves a return to this region.[38]

In the Middle Ages, the Church doctors taught that all human music bore a definite relation to the eternal and abstract music of universal order. Celestial sounds were produced by the turning of the glassy spheres. This prevailing view of heavenly music was complemented by another, the origin of which could be traced to a myth recorded in Plato's *Republic*. The myth spoke of the heavenly spheres as bearing, on each of their surfaces, a siren, "who goes round with them, hymning a single tone or note." Angels took the place of sirens and did the hymning in a Christianized universe.[39] But whether it was the great celestial instrument of the ancients, or the towering choir filled with angelic voices of the medievals, how strikingly different it is from the silent empty space of our time, intimations of which already alarmed Pascal in the seventeenth century.

To the Pythagoreans, harmony called to mind music and its mathematical intervals. In late antiquity, another meaning of harmony (also of Pythagorean inspiration) emerged, which too exerted an enduring influence on European thought: the concordance between the universal order and human lives. Harmony in heaven promoted harmony in the human soul. The music of the spheres had the power to assuage the savage breast and seed it with virtues. As the fourth-century philosopher Macrobius put it,

> Every soul in the world is allured by musical sounds so that not only those who are more refined in their habits, but all the barbarous peoples as well, have adopted songs by which they are inflamed with courage or wooed to pleasure; for the soul carries with it into the body a memory of the music which it knew in the sky, and is so captivated by its charm that there is no breast so cruel or savage as not to be gripped by the spell of such an appeal.[40]

In the past, music meant communal participatory performance; in our time, it assumes the presence of listeners, or even a single listener. Music is there in its own right, to be attended to in silent concentration for perhaps twenty minutes or longer at a stretch, either alone in a special chamber set aside in the house, or with others in public concert halls.

This radical change of attitude, which is unique to the West, occurred in the seventeenth century as part of growing disenchantment with the world, under pressure from scientific successes that significantly altered how educated Europeans perceived reality. The music of the spheres, despite its late affirmation by such noted astronomers as Kepler, fell silent. Yet music as a human creation not only did not fall from grace but actually gained importance. It came to be valued for itself. From the seventeenth to the nineteenth century, music could fairly be regarded as Europe's supreme achievement in art, offering more radical innovation and reaching more successfully to greater heights and depths of expression than did the sister arts of architecture, painting, and literature. During this period, two technical and conceptual changes—one, tempo and dynamics; the other, loudness and frequency range—set Western music sharply apart from that of other high cultures. Their joint effect was to expand markedly the European's sense of time and space—an expansion that was already emerging under the impetus of other forces. In the seventeenth century, as the concept of time in natural philosophy and the experience of time in daily life became more linear-directional and dynamic, music also tended to show more and more vigorous movement. A. C. Schuldt writes: "Not for this age was the serene flow of [Renaissance Mass or madrigal]. . . . No other period, no other culture, has produced a music so imbued with directional energy. Baroque music is goal oriented, driving always toward its triumphant cadences, its final, emphatic closure." This dynamic thrust, with its harmonic pressures and insistent rhythms, lasted well into the nineteenth century.[41]

The second highly distinctive feature of Western music, closely allied to the first, was the vast expansion of acoustic space, made possible by a growth of both intensity (loudness) and frequency range (pitch). The sound of a piano or an ensemble could be either so soft that one must move to the edge of the seat to catch it, or so unexpectedly loud—like the crash of thunder—as to make one almost jump out of one's seat. Pitch ranges from the deep rumble of drums that seems to wrap around one's body, to the soaring note of a violin that fades into the distant horizon. Perspectival space is thus manifest in post-Renaissance music as well as in Renaissance and post-Renaissance landscape painting, where it is far better known. A feeling for distance, which we tend to think of as an exclusive effect of visual-kinesthetic experience, is powerfully enhanced by sound: as Murray Schafer puts it, "the real space of the concert hall is extended in the virtual space of dynamics."[42]

Music written since 1600 demands concentrated listening; hence the observation of silence while it is played, the moment of stillness—a signal for attention and a mark of respect—before and after each piece, and the practice of playing music in its own chamber. A listener is essentially alone, even in a hall packed with people, absorbed in the unfolding, forward movement of sound and in the virtual acoustic space of foreground and distant horizon. Such music, even though it appears to offer a world "out there" rather than the enveloping, almost tactile sensuality of rhythms and beats, can profoundly move a human individual. As religion declines, music, at one time religion's servant, presents itself as a source of strength and comfort, if not of salvation. Beethoven's attitude illustrates the change. When mounting evidence forced him to face the terrifying prospect of deafness, he wrote: "Such incidents [of deafness] have driven me almost to despair; a little more . . . and I would have ended my life—it was only my art that held me back."[43] Testimonies to the deeply personal significance of music are numerous and heartfelt. Pablo Casals, at age ninety-three, said that music filled him "with awareness of the wonder of life, with a feeling of the incredible marvel of being a human being."[44] To Claude Lévi-Strauss, music is the supreme mystery of man's humanity.[45] Wittgenstein confessed to one of his few intimates that it was the slow movement in Brahms's Third Quartet that had pulled him back from the brink of suicide.[46]

PURE MUSIC AND COGNITION

Purely instrumental music, music alone, has existed in the West since at least the end of the sixteenth century, but not until the eighteenth was it accepted with full institutional support; and as the culmination of aesthetic experience—an experience often considered superior to, and of greater intensity than, that of all other contemporary art forms—pure music reached a pinnacle of prestige later still, in the age of Beethoven. What is this world of nonrepresentational, instrumental sound, seemingly remote from common human experience and yet capable of touching us at the deepest level? Peter Kivy has attempted an answer, and I shall draw upon his insights in the remaining pages of this chapter.[47]

Music affects us emotionally for a variety of reasons. We have an unreflexive, physiological response to certain rhythms, such as the martial beat of a military band, whether we want to or not. Music also affects us because it can seem meaningful, like language, even though it is wordless; it resembles human speech because its ordered periodic sound is quasi-syntactical. Music can have the emotional quality of human speech: the music's tone conveys passion just as a tone of voice is able to do even when the words are lost to the wind. We are biologically adapted to hear sounds as animate. By analogy and transference, even sounds made by plants (rustling leaves) and inanimate nature (wind or falling rock) may be heard as voices and cries conveying the emotions of alarm, pleasure, sadness, and melancholy. Pure music, like human voices and the sounds of nature, has the power to evoke these garden-variety emotions. But it can do far more. Gerald Brenan expresses his wonder thus:

> [In] listening to a sonata we are carried through a succession of complex emotions, enough to fill years of life, in the space of half an hour. The range, the precision and the delicacy of our re-

sponses surprise us, for we did not know before that we had such a rich capacity for feeling or such nimbleness in moving from one mood to another under the mastery of that magical spell. Yet so poor is our language to express states of mind that we can only use such vague terms as gaiety, melancholy, humour or pathos to describe what we have felt. . . . Music gives us a finer texture to our emotional responses than our lives can ever give us: it offers us, one could say, a foretaste of that celestial Utopia on which we would like to think that a perfected humanity could one day plant its flag. . . .[48]

Pure music's power seems mysterious. The orchestra is out there, the music it plays may extend space further into the horizon; and I am here. There is this distance between us, and yet the sounds produced by the orchestra can penetrate my innermost being, making me feel in extraordinary ways that disappear as soon as the music ends. Putting it thus, I appear to be saying that the music acts on me and that I, as listener, passively react. Yet reflection on the experience shows that the passivity is exaggerated, that I am not merely responding but responding with the mind, actively, that my pleasure is not merely visceral but "full of mind." And I would go further and argue, with Kivy, that I respond with the mind even when I am musically illiterate—even when I lack the technical vocabulary to say what is happening in the music. I can enjoy the music, be under its spell, and think about what I am enjoying at the same time, which is what having an aesthetic experience means. Aesthetic experience has as its essence this contradictoriness between passivity—this almost wired-in response—and active appreciation. "It is important and, I think, liberating," Kivy writes, "to perceive how deeply cognitive the musical experience really is. . . . A musical tone may seem to be some kind of paradigm of the simple, unanalyzably beautiful. I suggest it is nothing of the sort. A musical tone, the smallest particle of musical existence, is beautiful because it is *interesting*."[49]

Music lovers may well respond to music differently. Some choose to close their eyes and submit to the sonic flood. They want to be overwhelmed, to drown the sense of self—its particular location in space, its humiliating subjection to time—in the intensities and rhythms of sound. Others, more sophisticated, choose to attend to, say, a fugue's beauty of structure—to how Bach handles the stretto, how, starting on the seventh degree of the scale rather than the first, he gives the theme a whole new harmonic cast. A few extremists at the intellectualist end of the spectrum find aesthetic enjoyment almost wholly in the work's compositional intricacies. Nicolas Slonimsky, musicologist and mathematician, is one such individual. Slonimsky plays down the sensual aspect of musical experience. "Waves rising and falling, that sort of thing" reminds him of the Gulf of Finland, which used to make him seasick. "I don't see," he says, "what value there would be in experiencing music in that way." Lawrence Wechsler, describing this "boy wonder," points the contrast thus: "Leibniz once wrote, in one of his famous letters to Goldbach, 'Music is a hidden arithmetical activity of a mind that does not know it is counting.' But Slonimsky hears the counting. For him, the love of music is a mathematical passion—it is calculation taking flight, a structural transport."[50]

The world of sound affects us deeply for many reasons. Let me recapture three of them. One, the dynamism of sound is the dynamism of life: a soundless world is a dead world; two, the sound of human voices is so important to the acquisition of speech—to language and our status as human beings—that we tend to interpret the sounds of nature (both animate and inanimate) as quasi-human "voices"; three, whereas certain kinds of music produce almost tactile sensations (for example, the loud and deep boom of rock), others open up spaces "out there" and invite a more contemplative frame of mind. This last power of sound is one that is more commonly exhibited in the world of sight, to which we will now turn.

CHAPTER 5

Visual Delight and Splendor

S ight is valued above all the other senses. True, we can be persuaded that touch and hearing are more basic—the one to survival, the other to the acquisition of language. Nevertheless, sight enjoys primacy. It immediately gives us a world "out there." Self, without a world, is reduced to mere body. All senses give us a world, but the visual one has the greatest definition and scope. This expansive visual world is both sensual and intellectual. It is sensual, not only because of its colors and shapes, but also because of its tactile quality: we can almost feel what we see—smile with pleasure as we look at a fluffy blanket. It is intellectual because somehow to see is to think and to understand: sight is coupled with insight, and to exercise the mind is to see with "the mind's eye." Perhaps most important of all, the primacy of sight rests on simple experience. Open our eyes, and the world spreads before us in all its vividness and color; close them, and it is instantly wiped out and we are plunged in darkness. One moment, the world is an enticing space inviting us to enter; the next, it collapses to the limit of our body and we are helplessly disoriented.

As with the other senses, the visual world has both general characteristics shared by people everywhere and characteristics distinctive to sociocultural groups and individuals. But the particularities of visual experience seem to be more varied, the experiential differences between one group or individual and another greater. One of the many possible complex reasons for these differences may be the greater degree to which thought—the directed, intentional, and reflective thrust of the mind—is involved in the act of seeing, and the larger world that sight opens up. The smell of seaweed remains much the same in childhood and adulthood, but what people see at the seashore as adults may be significantly different from what they saw as children, even if they look in the same direction at physical features that have not themselves altered.

Composition versus Pattern

Of those visual experiences that are common to humankind, perhaps the most salient is reality as composition—a three-dimensional space defined by objects that are themselves three-dimensional. Objects, for humans, are things that can be picked up, extracted from their environment, and examined for their practical or inherent (aesthetic) interest. Our flexible and sensitive hands play a role in this experience. If we can *see* a ball as rounded and discrete, it is in part because we have had the experience of picking it up and feeling its shape and size. Nonprimate mammals cannot perform this simple act. Their paws cannot grasp a ball, only touch a small part of its surface and push it. In part for this reason, their visual world is one of pattern rather than of composition. Humans, then, live among objects located at various distances in space. Human eyes can see objects both close at hand, unlike other mammals, and at a great distance. Sight is our distance sensor, as smell is for many other animals.[1]

Most mammals judge objects to be recognizably the same over time by virtue of their consistent odor rather than through visual cues. A lion, for instance, cannot depend on eyes alone in hunting; the visual identity of prey is strongly affected by its context—its position within a pattern. When the prey moves and its context alters, the lion may no longer *see* it as the same thing, although the lion may continue to track it by smell. Pattern is, of course, also a pervasive component of human visual reality. We see pattern everywhere. Its discernment has practical value: a hunter traces his prey through an understanding of the pattern of the spores; a farmer prognosticates weather from the configuration of clouds in the sky. For humans, however, the pattern mode of seeing is not dictated merely by the physiology of their eyes; rather, it often appears as superior exercises in abstraction. One deliberately focuses on the bifurcation of the branches rather than on the branches individually; one deliberately attends to the pattern of the footprints rather than to the fact of shallow depressions made by human feet. Pattern can have, for humans, a strong aesthetic appeal: one pauses to admire the ripples in the sand or the stripes of a zebra without some economic purpose in mind. Decorative art, which is universally practiced and of the greatest antiquity, shows how strongly disposed people are to the orderly arrangement of things, especially when the order is one of simple repetition that can be grasped at a glance.[2]

Influences of
Environment and Culture

Aside from hallucination, which is always a possibility, people learn to see what is actually there—the natural and built environments that can vary enormously from one location to another. People have lived in mountains, where verticality prevails, and on vast plains ruled by horizontals; they have lived in exposed environments such as the savanna and the Arctic, and in enclosed, womblike places such as the tropical rain forest; they have lived in environments over-whelmingly green, where even the blue of the sky occurs only as broken patches in a green canopy, and in environments where the color green is a minor tincture in a landscape that, for long periods, is overwhelmingly white and gray; they have lived in the "carpentered" worlds of the city, where the straight line and the rectangle predominate, and in places where the rectilinear is seldom encountered, either in nature or in the architecture of round huts and circular corrals; they have lived in settings that constantly invite the eye to sweep over distant horizons and focus on distant objects, and in dense forests or cities, where the vista—the distant view—is no part of day-to-day experience.

Obviously, the environment constrains. It guides and puts limits on how the visual capacity develops. The eyes may be our distance sensor, but for the Mbuti of the Ituri forest there is no need or opportunity to exercise vision for that purpose, since trees block their sightlines everywhere and the edible plants and wild game needed for survival are exposed to view only when they are already within easy physical reach.[3] Biology imposes other constraints: some individuals are born visually handicapped; others enjoy unusual advantages. An extreme case is that of the twins John and Michael who, when a box of matches fell on the floor and the matches spilled out, immediately said, correctly, "I I I." They actually *saw*, at a glance, one hundred and eleven matches. Their exceptional ability highlights the limitations of normally sighted people.[4]

Culture also directs people's vision; it offers both constraint and empowerment. Parts of the Kalahari Desert seem utterly barren to the Western visitor, but to the locals—the Kung San (Bushmen)—they have ample resources and are by no means featureless. Much of the ethnographic literature reports visual acuity of a high order. For example, Claude Lévi-Strauss knew members of a tribe who claimed that they could see the planet Venus in full daylight. Astronomers confirmed to him that this feat was possible, given the amount of light the planet reflects. Lévi-Strauss also discovered in old navigation treatises that European sailors were at one time perfectly able to discern Venus in daylight.[5] It is for lack of training that we cannot do so now.

Sight guides every step of our practical life. Everything that comes into our field of vision is interpreted to enable us to navigate through space and do the sorts of things that need to be done. But the capacity that empowers us in practical life may also be a condition for aesthetic appreciation; the two spheres are not always separate and distinct. For instance, clarity of vision not only enables a hunter to see prey at a distance; it also fulfills an aesthetic role, by making an object vividly present. According to the art historian Bernard Berenson, "the trick [in painting vistas] is to make objects, and in particular human figures, no matter how diminished in size, as tangibly visible as if you could touch them. To identify them so clearly and with such detail as they increasingly recede from the eye gives one the feeling that at last one has got the right kind of world, where sight is not dimmed by distance—truly a life-enhancing experience."[6] Thus we can imagine the hunter having a life-enhancing experience as he follows his prey and anticipates success. He feels good because he can see so well, and this awareness of the vital presence of the world and its objects contains, inevitably, moments of aesthetic thrill and uplift.

Practical life and aesthetic experience intimately intermesh in another way—so commonplace that we hardly give it a thought—in our natural inclination to do our tasks, whatever they are, with efficiency and skill. When I vacuum the carpet and create neat swathes of flattened fibers, when I look at a cleanly typed page, when the plowman strives to produce a straight furrow or the carpenter looks at the joints in his woodwork with a sense of pride, there is necessarily an aesthetic tinge to the satisfaction. All these activities are attempts to maintain or create small fields of order and meaning, temporary stays against fuzziness and chaos, which can be viewed, however fleetingly, with the pleasure of an artist.

We are more sensitive to the myriad small executions and offerings of beauty than we realize. One reason for the widespread view that most people are practical rather than aesthetic is that "practical" has come to mean almost the opposite of "aesthetic." People in modern times take for granted that the primary purpose of the eye is efficiency in practical matters rather than contemplation of beauty or God, as medieval theologians appear to have believed. A church steeple may be a thing of beauty, but I am aware of it only as the landmark that tells me to turn right on my way to the office. A pebble that I have picked up on the beach for its appealing color and shape now lies unnoticed on my desk as a paperweight. And so on. Such examples, which can be multiplied ad infinitum, say how rarely we attend to the world aesthetically. And yet we do attend to the world aesthetically, not often (it is true) as an art student does in an art gallery or as a nature lover does on the seashore, but often nevertheless, as alert individuals do, glancingly, during the pauses and among the interstices of practical life.

Culture sets general and specific goals, the attainment of which may call for the acute development of certain sensory capabilities. Culture also tells people what is beautiful. Again, we can see this as constraint or as empowerment, or as both on the grounds that one's field of operation must be delimited for perception and action to be effective. One result of culture's function in setting standards is that not everyone shares our sense of what is beautiful. One people's architectural pride can leave another people quite unimpressed. Thus Pacific Islanders, who are strangers to the silhouette of a modern city, may deem it unworthy of aesthetic contemplation. It is understandable that people favor what they have made themselves. But people differ even in the extent to which they admire natural environments, such as the dramatic play of color and form in a brilliant sunset. The anthropologist Raymond Firth recalls an exceptionally beautiful twilight scene at Tikopia, a Polynesian island in the South Pacific. "The cloud hid so completely the setting sun that it brought dusk before the sunset. Then just when the sun was on the point of setting, it broke free, and . . . sent lurid crimson colour on the sea, the walls of houses and the trunks of trees, while the land had begun to take on the shades of darkness." This "weird conjunction of dusky shadows and red sunset light," Firth comments, "made even the natives pause to stare, but they assigned it no special significance."

They were struck momentarily by the scene's strangeness and excess rather than by its beauty. "As a rule the more subtle and really more beautiful differentiation of shades escapes their notice."[7]

Color

Another general property of human perception, shared with other primates, is color vision. We tend to take it for granted. One way to begin to understand the importance of color to the tenor of life is to imagine a world with a color scheme different from the one we are used to: an environment of enduring green, as distinct from one in which foliage offers a chromatic extravaganza of seasonal changes; or the clarity and brilliance of the desert (the bright hues of sand dunes at twilight, or sparkling stars against the velvet black of the sky), as distinct from the subtle lights and shades, the misty wash of subdued colors, in the humid northern climes of Europe. One student at the University of Minnesota who flew out to California in early April, when the Upper Midwest was still dull gray and brown under the cover of melting snow, found the experience like shifting from black-and-white to color television.

Color vision varies from individual to individual. The colorblind are clearly handicapped by the inconvenience of being unable to tell the difference between red and green in traffic lights. But we are all "colorblind" to varying degrees; some people are more sensitive than others to certain colors, finding it easier to pick out, say, the reds and greens in gloomy light. However, most of us are oblivious to the relative handicaps and strengths in our color vision, unless we happen to be artists or are in the design profession. We are content with the capabilities we have. It takes an extreme case to shake us out of the torpor of habitude and make us aware of the importance of color vision not only in terms of practical efficiency, but also in terms of happiness—the vital sense of life, aesthesia.

Such an extreme case is presented by neurologist Oliver Sacks and Robert Wasserman in their account of a sixty-five-year-old painter, Jonathan I., who became totally colorblind as the result of a car accident. Jonathan did not realize at first what had happened to him.

As he drove to work, he thought the day strangely misty and bleached even though he could see the sun shining. The first true shock came when he opened the door to his studio and found his paintings, which were brilliantly colored abstractions, devoid of color—shades of gray; and without color, he noted to his despair, meaning drained out of the works. Loss of color vision may not immediately strike us as so horrible because we think the world then turns into the sort of sharply contrasting images that we associate with good black-and-white photographs and movies. But the colorless world that Jonathan plunged into was not like that. It looked not clean-cut and dramatic, but "dirty." The whites were "glaring, yet discolored and off-white, the blacks cavernous—everything wrong, unnatural, stained, impure." Human skin looked "rat-colored," which put him off sex. Foods were disgusting in their grayish, dead appearance. Flowers could be distinguished only by their shape and smell, and what is spring without its brilliant colors? Clouds, in their off-whiteness, could scarcely be separated from the bleached "blue" of the sky. As for the rainbow, a symbol of hope, Jonathan could see it "only as a colorless semicircle in the sky."[8]

Jonathan's misfortune shows how important color is to happiness. Life is often described as either "dull" or "colorful." We now realize that even the most philistine person, to enjoy the world, must have a sensitivity to color. People are more aesthetically inclined than they know or are willing to admit. However, individuals and cultures differ widely in their capacity for and style of appreciation. Age, predictably, makes a difference. Young children are drawn to vivid primary colors. A wider range of color awareness and a greater liking for the subtler tones come with age and education. As for the preferences of different cultures, Brent Berlin and Paul Kay note that all cultures have three basic color terms—black, white, and red—and that the more advanced the cultures are materially and politically the more terms they possess; and this means that among the riches a culture gains as it becomes more complexly structured is a more brightly and subtly colored world.[9]

In any populous and advanced society, some fortunate individuals—for a variety of reasons—show exceptional sensitivity to color.

Consider Augustine of Hippo and Simone de Beauvoir, separated in time by 1,500 years and in outlook by unbridgeable differences. For Augustine, nature had profound appeal. He loved North Africa's sunlight, which he called the "Queen of Colors." He sat under it, bathing in its limpid beauty and regretting that he must eventually go indoors. "I miss it; and if I am long deprived of it, I grow depressed." To the old bishop, the beauty of the wide bay of Hippo was a foretaste of heaven. "There is the grandeur of the spectacle of the sea itself, as it slips on and off its many colors like robes, and now is all shades of green, now purple, now sky-blue. . . . And all these are here consolations for us, for us unhappy, punished men: these are not the rewards of the blessed. What can these be like, then, if such things here are so many, so great, and of such quality?"[10]

Simone de Beauvoir found the Sahara Desert "as alive as the sea. . . . the tints of the dunes changed according to the time of day and the angle of the light: golden as apricots from far off, when we drove close to them they turned to freshly made butter; behind us they grew pink; from sand to rock, the materials of which the desert was made varied as much as its tints; sinuous or sharp-edged, their forms produced an infinity of modulations within the deceptive monotony of the erg."[11] Although de Beauvoir's travel writing follows an established tradition, such descriptions are more than mere literary habit. Clearly, natural beauty touched her to the quick, as it did Augustine. Nevertheless, time and culture do make a difference. What Augustine saw in the changing sunlight and sea was transparent to another realm not of this world. For Simone de Beauvoir, a secular person living in a secular age, the colors of the desert, for all their loveliness, bore no intimation of immortality. There was no Beyond for her that could give a transcendent meaning to what she saw and recorded.

Gemlike Fire versus Twilight

Individuals differ as a consequence of temperament and disposition. Group differences reflect the norms of culture, which change over time. The changes may be superficial, as in the rapid shifts of fashion in the modern world, or profound, reflecting radical alterations in cultural practice and world view. Consider the contrast between the mineral or gemlike glitter of the Middle Ages with the phantasmagorical, twilight colors of nineteenth-century Romanticism. If a time machine could take us to the Middle Ages, we might be shocked, first, by its pungent odors, then by the intimate commingling of wealth and poverty, youthful vigor and disease. But we might also be shocked—or pleasantly surprised—by its brilliant colors. The chronicler Jean Froissart (1333?–1400) was too busy with his narratives to pause and remark on beauty. Nevertheless, Johan Huizinga observes, "one or two spectacles . . . never fail to enrapture him: that of vessels on the sea with their pavilions and streamers, with their rich decoration of many-colored blazons, sparkling in the sunshine; or the play of reflected sunlight on the helmets and cuirasses, on the points of the lances, the gay colors of the pennons and banners, of a troop of cavaliers on the march." Garments, normally colorful, became gaudy on festive occasions. Every combination of colors was allowed: red with blue, blue with violet. One lady appeared at a banquet in "violet-colored silk on a hackney covered with a housing of blue silk, led by three men in vermilion-tinted silk and in hoods of green silk."[12] Gems, sewn on the robes of lords and ladies, sparkled in the sun and glowed under candlelight. The cathedral itself was a vast jewelbox, radiant with the light that poured through its rose windows.[13]

In the nineteenth century, gems were no longer conspicuously loaded on bosoms and costumes, and the facades of public buildings had long ceased to project the vivid colors of the past. Somberness of tone prevailed in cityscapes, and nowhere was this more evident than in the black attire, worn like a uniform, by the male members of the bourgeoisie.[14] As interest in bright primary colors and mineral glitter declined, a new palette gained favor—the colors of twilight, dark hues suffused or seamed with red and orange, shading into yellow and the pale silver of the moon. The Industrial Revolution played an important role here. Flame and smoke from the factories and furnaces, working late into the night, gave a lurid cast to the evening sky. The horror that initially greeted this desecration of nature and the countryside was transmuted, among artists and writers, into the aesthetic concept of the sublime: the threatening, the garish, and the deathly were seen to have their own infernal grandeur. "They raised their eyes, and saw a lurid glare hanging in the dark sky; the dull reflection of some distant fire."[15] And yet Little Nell and her grandfather found warmth and shelter in the factory that produced the distant fire. If Dickens's portrayal of industrial landscapes was not totally condemnatory, even less so were the portrayals of late eighteenth- and early nineteenth-century painters, who could not help but show on their canvases their enchantment with the newly discovered dramatic colors. Examples are Joseph Wright's *Blast Furnace by Moonlight*, George Robertson's portrayal of the Nant-y-Glo iron works (also shown against moonlight), and P. J. de Loutherberg's *Coalbrookdale by Night*. Even John Martin's *The Great Day of His Wrath* might be included in the list, and this despite the fact that it was intended to show the Black Country in the unsympathetic light of an apocalypse.[16]

There was another aspect to twilight—quiescent and touched by melancholy. Compared with the pearly glow of dawn, which became a stock Romantic image of hope that appealed to calendar art, the dimming horizon of dusk, the evening of life, and the Twilight of the Gods were treated by artists and writers with greater subtlety and seriousness. Carmine-and-ocher hues that represented the infernal energies of hell and industrial furnace could, with slight changes of color scheme and of perspective, be made to portray the reflective mood of a dying day or world, or, possibly, of a lost world of forgotten purity and silence, as in Frederick Church's painting of subdued exaltation, *Twilight in the Wilderness*.[17] Twilight colors were not confined to depictions of nature, or, in more somber tones, to depictions of the industrialized landscape. They also found favor in the domestic sphere. Upper-middle-class homeowners in Victorian and Edwardian England and Anglo-America showed a certain fondness for the glow of evening firelight, as well as for the iridescent shifting palette of Roman glass and Persian tiles.[18]

Intense and Joyful Seeing

All humans value sight, which they need for the efficient conduct of day-to-day affairs, but also to enjoy the world's visible presence.

Nevertheless, striking differences occur. Some individuals live as though suffering from a perpetual head cold: everything is seen hazily; objects lack distinctiveness, trees and houses appear as generic types like those drawn by young children.[19] At the other extreme are artists, who show an unusual absorption in the quiddity of things. Gustave Flaubert claimed to have almost voluptuous sensations simply from seeing well. "Often, à propos of no matter what, a drop of water, a shell, a hair, you stopped and stayed motionless, eyes fixed, heart open. . . . The object you contemplated seemed to encroach upon you, by as much as you inclined yourself toward it, and bonds were established. . . . Sometimes by dint of gazing at a pebble, an animal, a picture, I felt myself enter into them. Communications between human beings are not more intense."[20] Eugène Delacroix is famed for his large epochal canvases, yet his diaries reveal a keen appreciation of the most humble objects—the garden slug, for instance. Of its coloration, he noted that it was "spotted like a jaguar, with broad rings upon its back and sides, turning into single spots on the head and near the stomach, where it was lighter in tone, as in quadrupeds."[21] The slug, yes, we might say. Nature is somehow always significant. But what about ordinary utilitarian objects? The poet Gerard Manley Hopkins, as a student in a Jesuit college, was assigned the task of cleaning the outdoor water closets. During a cold snap, he noted that "the slate slabs of the urinal . . . are frosted in graceful sprays."[22]

Artists, with their intense awareness of the mysterious beauty of the world, may want to return to a particular enthrallment and try to capture its quality again and again. Examples include Hokusai's multiple depictions of Mount Fuji, Cézanne's similar efforts with regard to Mount St.-Victoire, and especially Monet's remarkable series *Mornings on the Seine*, executed in 1896–97. The composition of the nine members of the series is virtually identical, and the pictures create a dramatic mood "without a touch of anecdote—no figure, no boat, no individualized tree." What Monet wants to show is not so much a composition of bounded objects as something far more elusive—the morning sky and air that cast a subtle, changing veil over the water and the land. "In no two of the series is the atmosphere the same; in all of the eight . . . the tonal modulations of intervening moisture-laden air, each matched by a subdued reflection in the surface of the Seine, are magisterial."[23]

In the intensity of the gaze, the self disappears—or rather, it contracts into the abstraction of an eye. Cézanne described Monet as "only an eye—but what an eye!" There is selflessness in this absorption. The object becomes so real that the self, by comparison, feels insubstantial. What really matters is not the self (certainly not the social, anxious self) or even the self as all-seeing eye, but rather the empirical reality out there. We can have the sensibility if not the skill of artists, if we know something of the facticity and "weight" of objects. "The cornfields of the Midwest," notes Suzannah Lessard, "are so supremely factual that they seem to discredit everything unempirical. The landscape made even the life inside the bus seem phantasmagorical, as if the bus were a head, the windows were the eyes, and the people in the bus were thoughts."[24]

Visual splendor can enable people to forget (momentarily) even the harshest conditions of life. Consider this account of a German concentration camp by Viktor Frankl:

> One evening, when we were already resting on the floor of our hut, dead tired, soup bowls in hand, a fellow prisoner rushed in and asked us to run out to the assembly grounds and see the wonderful sunset. Standing outside we saw sinister clouds glowing in the west and the whole sky alive with clouds of ever changing shapes and colors, from steel blue to blood red. The desolate gray mud huts provided a sharp contrast, while the puddles on the muddy ground reflected the glowing sky. Then, after minutes of moving silence, one prisoner said to another, "How beautiful the world *could* be."[25]

The Beauty of Order:
Cosmos and Microcosm

Sunsets and sunrises are not universally admired. It is only at a certain stage of cultural development and under certain conditions that a people may look admiringly at the coloring of the early evening or morning sky. Whether delight in an aspect of the world is restricted to a particular culture or widely shared seems to depend, to some extent, on scale. At a microscale, people in different parts of the world may like the same sorts of things: a shapely pebble on the beach, the smile of a child, the texture of polished wood are common human loves rather than the acquired response of a group. At the medium scale of sunset, seashore, mountain, and other picturesque landscapes, culture does indeed exert a powerful influence: thus, for historical and cultural reasons, in some societies people like mountains; in others, they don't. Beyond the medium scale, nature and landscape may again evoke transcultural admiration, all the more so if attention is directed to their abstract qualities: the appeal of order manifest in the movement of the heavenly bodies, in the alignment of fields and farms, and in the plan and architecture of the city. *Cosmos*, *countryside*, and *city* are familiar ideal-types.

Sun and moon, by their commanding presence and by their real or imagined impact on terrestrial nature, entered human consciousness in the earliest religions to move beyond the worship of the chthonian deities. Later, the stars were also carefully observed, for they, too, were believed to control human fate. Although agricultural civilizations noted the movement of the stars and the rhythm of the seasons for practical reasons, the splendid rituals they created to mark critical points in the annual cycle are testimony that they also appreciated its cosmic grandeur. The ancient Greeks were unique in admiring the heavenly bodies for themselves, and not because they impinged on human affairs. "To what end are you in the world?" This question, a sort of catechism for the Greeks, provoked in Anaxagoras the answer, "To behold sun, moon, and sky."[26] The stars were venerated and loved by the philosophers for "their paradigmatic existence," as Hans Jonas puts it. "The purity of their substance, the perfection of their circular motion, the unimpededness with which in thus moving they follow their own law, the incorruptibility of their being and the immutability of their courses"—all these attributes made them seem "divine," not as gods and goddesses were divine, but in an altogether more elevated, suprapersonal sense.[27] Early Christianity retained something of this attitude. "What is the primary value of sight?" The appropriate reply to this question in Christian catechism was: "To beget philosophy, for no man would seek God nor aspire to piety unless he had first seen the sky and the stars." God gave men eyes, Chalcidius wrote, in order that they might observe "the wheeling movements of mind and providence in the sky" and try to imitate as best they could that wisdom, serenity, and peace.[28]

Compared with the lawful courses of the stars, terrestrial nature seemed a threatening chaos to almost all advanced (that is, plow-using) agricultural peoples. Carved out of that chaos or wilderness were the farm and its human dwellings. To this transformed world, always threatened by the instabilities of nature, people owed their allegiance. The sentiment of piety was no doubt mixed with the pride of possession, but along with these sentiments, and fully congruent with them, went also an awareness, however subdued and inarticulate, of the comeliness of the farm—its display of furrowed fields, fruit trees in rows, and buildings. Already, in the famous description of the Shield of Achilles in Homer's *Iliad*, an appreciation of the farm scene has a strong aesthetic component:

> *There too he sculptured a broad fallow field*
> *Of soft rich mould, thrice ploughed, and over which*
> *Walked many a ploughman, guiding to and fro*
> *His steers, and when on their return they reached*
> *The border of the field the master came*
> *To meet them, placing in the hands of each*
> *A goblet of rich wine. Then turned they back*
> *Along the furrows, diligent to reach*
> *Their distant end. All dark behind the plough*
> *The ridges lay, a marvel to the sight. . . .*[29]

"A marvel to the sight." Thus the plowmen could appraise their handiwork, and the master too, who brought them the wine. The farm is the "middle landscape" between wilderness and the city. Virgil saw the fertile Po plain—its ancient beeches and dark oaks standing among the pasturelands, with little herds of sheep and goat grazing among them—as the middle landscape between the Alps and Rome, an area of precarious achievement threatened by the violence of nature and barbarism on the one side and by the excesses of civilization on the other.[30] Whenever cities reached a certain size and complexity, sentiment among their well-to-do citizens tended to swing in favor of the farm and of country living. This has happened repeatedly in Europe—in the Hellenistic Age, the Augustan Age, the Renaissance, and the eighteenth century. And in China too. Taoism was a nature philosophy that allowed Confucian scholar-officials, unsuccessful or too harassed at court, to withdraw to their native village, where a Tao Yuan-ming could write in the fourth century (and here I paraphrase his poem): "In early summer the woods and herbs are thriving. Branches sway around my cottage, casting shade over it. As the numerous birds love their sanctuaries, so I love my cottage. After I have plowed and sown, I return to it to read my books."[31] The natural world of the Taoist is never truly wild: the human figure is always somewhere—travelers on a mountain path, the scholar-hermit in his cottage, and neat rice plots. In the United States, the middle landscape of family farms has been elevated to a mythic ideal. The farm, with its tidy fields and orchards, stands for wholesome living and political virtue, in contrast to the large commercial estate, which is commonly regarded as an arrogant intrusion of city wealth and technology into nature.[32]

The heavens display the exemplary motion of the stars. On earth, humans must create their own order: modest, as in the neatly plowed fields of the countryside; heroic, as in the geometric city. The great cities of antiquity were sacred spaces, ritual power houses, and ceremonial centers.[33] The human passion for order demands clear visual expression, and there is nothing clearer, and at the same time more laden with symbolic meaning, than the circle, the square, and the polygon. These forms occur rarely in nature. Apart from the sun, the full moon, and the rainbow, circularity is more commonly inferred than perceived—from the motion of the stars around Polaris, the trajectory of the sun, and the cycle of the seasons. Circular cities, such as the ancient Hittite city of Cincirli and Old Baghdad, capital of the Abbasid Caliphs, are rare, if only because of the difficulty of constructing circular walls of wide girth. Holy cities described as a mandala of concentric circles, such as the Hindu holy city of Banares, are only symbolically circular, not literally so. Rectilinear walled cities, in contrast, are common, and in the case of many ancient civilizations these shapes stood for the circular processes of the cosmos. The square shape of the traditional Chinese capital city, for instance, is a cosmic diagram, with the twelve gates, three on each side, representing the twelve months of the year.[34]

The cosmic city makes constant references to its transcendental nature. Thus the city's center is also, symbolically, its summit: in moving from the outer wall to the center one moves not only inward but upward. "Many of the sacred centers of Asia," writes Diana Eck, "have explicitly duplicated Mt. Meru [the mountain at the center of traditional Indian cosmology] in their structure. One of the most famous is the Khmer capital . . . now known as Ankor Thom, with its central temple-mountain of Bayon. Its high central tower, with five high terraces and towers, duplicates Meru, with its center and four directions." The link to the transcendent may also be enforced by myth. Thus Banares is said to be the place where "Siva's awesome shaft of light broke open the earth and shot up through the sky to pierce the very roof of heaven."[35] In the West, the Baroque plan of avenues radiating outward from central palace, statue, or monument (as in the case of Versailles or Washington, D.C.) makes obvious reference to the sun casting its rays over the human world.

The plans of cities with sacred or cosmic pretensions are readily grasped both visually and conceptually. Other cities offer more complex spatial-visual experience. Oxford, "a city of dreaming spires," though not a sacred place in the old sacerdotal priest-king sense, nevertheless has mythopoeic and heavenly aspirations. Its plan, rather than a display of simple geometric shapes, is one of surprises, a "fluid" order that has to be experienced in succession, as one wanders down a path. Thomas Sharp, Oxford's planner in the immediate post–World War II period, describes the kinds of experience a visitor may have.

> As he approaches the Bodleian from the top of Catte Street, there is nothing to be seen but its noble cube. Advancing, he sees first the rotunda, then the spire of St Mary's, then the dome of the Radcliffe coming into view. As this vast circular bulk separates from the bulk of the Bodleian, the tower of St Mary's also emerges. Despite the fact that each of the three buildings is in its own way as sophisticated a piece of architecture as there is, the ex-

perience is elemental, beyond the power of words or photographs to describe. Cube, cylinder, and cone are suddenly juxtaposed, or rather suddenly deploy the one from the other, with a result that is . . . sensational.[36]

Home and Sublime Nature

A universal human need is familiarity, at the back of which is routine. And the place where we are mostly likely to find a comfortable routine, and where we have first known it, is the home. At first we are likely to think that nothing is more contrary to the aesthetic, for whatever becomes routine fades from consciousness. Moreover, the necessities of life, which routine and order serve, tend to be viewed as outside the aesthetic realm. Two other traits of home further dissociate it from the aesthetic as commonly understood: the informal character of the order there and the typically diffuse multisensory ambience.

Nevertheless, aesthetic experiences are a commonplace of the home. There is, for a start, skill—the economy of means to achieve certain ends. Even the most ordinary household chores and social "ballets" have a curious fascination for passersby who, strolling along the sidewalk on a summer evening, happen to look into a window or open door. The aesthetic standing of the home may also be suspect because it is not conspicuously visual; and we have noted the strong visual bias in the Western world's conception of the beautiful. Of course the home offers visual beauty—from the bright copper pot to the landscape painting over the walnut cabinet; and it offers auditory delights, from the voice of the child reading a story aloud to recorded chamber music. But typically all one's senses are soothed or stimulated in some way: the fragrance of garlic in the stew, the texture of the armrest, the bells of an ice-cream van, the color of twilight. None, alone, is worthy of rapture, but together they make one feel good to be alive; they give savor to life.

The farther removed we are from home, the more our engagement with the environment tends to be conscious and visual rather than subconscious and multisensorial. The farm beyond the homestead is still part of the familiar world, and yet even in Homer's time the plowman could look back appreciatively at the deep furrows awash in the glow of the setting sun. An embryonic idea of landscape already existed then. Perhaps landscape—this visual composition of strong aesthetic appeal—is more common to human experience than we think. Rarer than the experience of beauty in landscape is its translation into a tangible work of art.

Beyond the orderly farm—the humanized landscape—lie dark forests, wild mountains, and stormy seas. These primal forces of nature in turn became landscape—objects of contemplation and portrayal that, in the eighteenth century, were designated as sublime. The meaning of sublime had changed over time. In antiquity Longinus, who invented the concept, applied it to the stable course of the stars as well as to the more unruly mountain and ocean. In the eighteenth and nineteenth centuries, Romanticism favored the wilder images of nature on the grounds that these evoked deeper and grander passions. But the sublime, as a type of human experience, is not merely an invention of philosopher-artists during a certain historical period. Whenever people step outside the protective enclosure of their known world, they risk encounter with some large, threatening force that yet holds an inexplicable attraction. One can be drawn to the sublime as, in a more religious phase of human history, one was drawn to the holy, to light, splendor and the numinous—the *mysterium tremendum* that is beyond human rational understanding.[37] Surely Job experienced the sublime when God spoke to him out of the whirlwind, as have mystics throughout the world who have pushed beyond normal experience to touch the heights and depths of the wholly other.

Since the eighteenth century, a new way of relating to reality has emerged, though differing from the old way only in degree. The essential difference is one of greater distancing. The new experience is more secular, less wrapped in dread of the sacred, more a deliberate effort to seek out, less an impingement from an external source (God or nature). In short, it is more aesthetic. Yet the overpowering transcendental element in the sublime remains. One contemplates beauty, even enjoys it. Contemplation implies a certain distance between the self and the other; enjoyment, for its part, implies the self's mastery of the other. Neither term is quite right for the experiencing of the sublime. Contemplation is more nearly correct: we do, after all, speak of contemplating the stars. What that word lacks is a vertiginous sense of being overcome and a premonition of danger that is also a lure.

Ice

In the modern period, only the great ice plateaus remain almost wholly free of people and their heavy imprint. There, experiencing the sublime is still an ever-present possibility. Ice plateaus lie at the other extremity of the world of shelter and nurture that is home. Histories of polar exploration and biographies of explorers show very mixed motives for going there. Before the eighteenth century, the chief motive appears to have been economic. Explorers hoped to find a way to the land of spices over the roof of the world. The myth of the Open Polar Sea made such attempts seem not unreasonable.[38] By the end of the eighteenth century, however, the idea that a route of commercial value could exist over the frozen Arctic had to be given up. From then on the most often publicly declared reason was science. Geography must be served. The rhetoric of science was not wholly convincing even to those who made it. Other motives were clearly at work, including the desire for adventure, to set a new record, acquire sufficient wealth to be independent, win glory for oneself and one's country, test the limit of human endurance, and something rarely stated in the open—a yearning for sublime experience, the loss of pedestrian consciousness in the vastness of nature, which is a kind of death.[39]

Of all the polar explorers, the Norwegian Fridtjof Nansen (1861–1930) and the American Richard E. Byrd (1888–1957) were perhaps the most introspective and philosophical. They left behind not only scientific observations but also reflections on nature, cosmos, and the meaning of life. Both believed that life was more likely to yield its deepest meaning in the silence, beauty, and terror of ice than in the quiet of a library.

Even as a boy, Nansen was enchanted by nature's spectacular

beauty. The aurora, though a familiar sight in his part of the world, never lost its power to impress. "However often we see this weird play of light," he wrote, "we never tire of gazing at it; it seems to cast a spell over both sight and sense till it is impossible to tear one's self away." Norse mythology contributed to the spell. "Is it the fire-giant Surt himself, striking his mighty silver harp, so that the strings tremble and sparkle in the glow of the flames of Muspellsheim?"[40] But even stronger evidence of Nansen's romantic temperament occurs in those passages of his journal where he merely recorded an event. In 1888, Nansen and his five companions were struggling to cross the Greenland ice cap. They put up sails on their sledges to take advantage of the wind. "It was rapidly getting dark, but the full moon was now rising, and she gave us light enough to see and avoid the worst crevasses. It was a curious sight for me to see the two vessels coming rushing along behind me, with their square Viking-like sails showing dark against the white snowfield and the big round disc of the moon behind."[41]

Nansen doubted the existence of God and did not believe in an afterlife. If life had a purpose, it was to use one's talents for the benefit of humankind. A successful organizer of Arctic expeditions, he also became an effective humanitarian on both a national and an international scale. Outwardly successful in every way, Nansen nevertheless suffered from depression in those periods when he was not harnessed to the strenuous demands of a polar exploration; and even in his preferred world of ice, where he saw beauty and splendor he often also saw death, as the first sentence of his two-volume work *Farthest North* attests: "Unseen and untrodden under their spotless mantle of ice the rigid polar regions slept the profound sleep of death from the earliest dawn of time."[42] The polar region was, for Nansen, the "kingdom" of death. Time itself seemed frozen. "Years come and go unnoticed in this world of ice. . . . In this silent nature no events ever happen. . . . There is nothing to view save the twinkling stars, immeasurably far away in the freezing night, and the flickering sheen of the aurora borealis. I can just discern close by the vague outline of the *Fram*, dimly standing out in the desolate gloom. . . .

Like an infinitesimal speck, the vessel seems lost amidst the boundless expanse of this realm of death."[43]

Richard Byrd was born in the year Fridtjof Nansen sailed across Greenland's ice cap. They lived thus a generation apart. The one became an admiral, the other a highly respected statesman. Both were successful men of the world and great polar explorers. What distinguished them from other explorers in both their own and earlier times was the way they turned their geographic voyages into voyages of self-discovery. In 1934, Byrd spent four and a half months alone on the Ross Ice Shelf of Antarctica. Science was his official justification; but a deeper reason was to discover for himself just how good peace and solitude could be in an environment that offered so little ordinary human comfort and consolation. Byrd did not find any easy answer. In a world in which no matter where he looked, "north, east, south, or west, the vista was the same, a spread of ice fanning to meet the horizon," he knew intense homesickness and despair; he came to think his stated scientific purpose a delusion and his desire for enlightenment a "dead-end street."[44]

Byrd enjoyed a more sanguine temperament than Nansen; yet from time to time a mournful tone also infected his prose. Of the icebergs enveloped in fog, he wrote: "Everywhere those stricken fleets of ice, bigger by far than all the navies in the world, [wandered] hopelessly through a smoking gloom." At his Ross Shelf camp he observed: "Even at midday the sun is only several times its diameter above the horizon. It is cold and dull. At its brightest it scarcely gives light enough to throw a shadow. A funereal gloom hangs in the twilight sky. This is the period between life and death. This is the way the world will look to the last man when it dies." More often, however, the entries in his journal affirm a sense of oneness with the cosmos. "The day was dying, the night being born—but with great peace. Here were the imponderable processes and forces of the cosmos, harmonious and soundless. Harmony, that was it! That was what came out of the silence. . . ."[45]

Both Nansen and Byrd, in their reports and diaries, expressed sentimental attachment to home. Hibernating with a mate in a primitive hut on Franz Josef Land, Nansen thought of his wife and daughter at home: "There she sits in the winter's evening sewing by lamplight. Beside her stands a young girl with blue eyes and golden hair playing with a doll. . . ."[46] Byrd, in one of his moments of despair at the Antarctic camp, declared: "At the end only two things really matter to a man, regardless of who he is: and they are the affection and understanding of his family."[47] Yet Nansen and Byrd freely left home, repeatedly, for something they could never quite explain to themselves. Home offers warmth, familiarity, and comfort. Beyond it is landscape, appreciated for its resources or its pleasing visual qualities. Still farther away is alien space. Confronted by its immensity and power, one cannot simply stand to the side and appraise it as one can before a rural scene amid rolling hills. Conflicting emotions, including fear, are aroused. The boundaries of self are threatened. Whereas absorption into the sensory realities of home means life, the loss of self in alien space—in the world of ice—can mean death. Travelers to the solitary regions of permanent ice appear to be half in love with a dangerous and piercing beauty and half in love with oblivion.

Whereas love of home is universal, the willingness to risk life itself for experiences of the sublime in places both remote and desolate is unique to a small number of individuals in Western society and culture. As we explore the capacities of the senses, we quickly realize that culture must be taken into account. However similar our basic physiological endowment, culture introduces remarkable diversities in the ways we perceive the world. In Part III, I will present a cultural-aesthetic spectrum that hints at the range of the differences.

III

A Cultural-Aesthetic Spectrum

Australian Aborigines, the Chinese,

and Medieval Europeans

*T*he power of the human senses to organize the world takes diverse forms, shaped by the larger cultures in which they operate. This chapter and the next detail the aesthetic qualities at the core of Australian Aboriginal, Chinese, medieval European, and modern American cultures. Far apart in time and space, these four worlds could hardly have less in common, or so we might think. Yet all possess an aesthetic-moral aspect—as revealed by their drive toward significance and form—and all demonstrate the power of the imagination to transcend group values held at a certain time by incorporating values from another group and thereby grow. We may be emotionally disposed to favor our own cultural ways because these are often also ingrained physical habits. But we may well develop an intellectual allegiance to whatever culture we deem liberal and liberating enough to remind us that, as individuals, we can surpass group norms.

Aboriginal Art
and Dreamtime

Whereas early European artists often depicted Australian Aborigines as Noble Savages, most early European settlers viewed them with contempt as they noted this dark-skinned people's lack of agriculture, permanent settlement, and material possessions.[1] However, those who knew something about ceremonial cultures in other parts of the world observed that Aboriginal dances, mimicry, and interpretative dramas accompanied by songs were comparable. As settlers moved inland, they discovered rock engravings and cave paintings whose quality aroused their curiosity and commanded their admiration. In 1876 R. Brough Smyth summed up knowledge of Aboriginal art at that time: "the practice of ornamenting caves, rocks, and trees, and cutting figures on the ground by removing grass is characteristic of this people. Their pictures are found in every part of the continent." Offering several descriptions "illustrative of their love of art," he urged his readers "to regard with a higher interest the first attempts of a savage people to imitate the forms of natural objects . . . and [of] incidents in their lives."[2]

The displays of this art vary predictably in both quantity and quality, reflecting variations in the natural environment. In general, the richer the environment, the denser the human population, and the more ceremonial exchanges and commerce that occur there, the more likely it is to have a wealth of art. The two richest areas are Arnhem Land and the Kimberley region of northern Australia. By contrast, the desert areas contain little artwork, if only for lack of surfaces—cave walls, suitable outcrops, or bark—to work on. Isolation and the need to move about frequently in search of food and water further militate against artistic production and innovation. Nevertheless, the aesthetic impulse appears even in the harshest settings. The Lake Eyre region, for example, is covered with shifting sand and sand hills, with no suitable rock faces for engraving and painting; periodic floods and erratic rain force the tribes to move about in small groups. Yet the local people erect brightly painted signposts (*toas*) that are far more than narrowly utilitarian: they not only indicate the destinations from a camp, but also portray the journeys of ancestral heroes.[3] Similarly, in the Great Victoria Desert, where water dominates the inhabitants' consciousness, waterholes are a common design motif on sacred boards and posts.[4]

Early European settlers generally described the life of the Aborigines as almost unbearably harsh. Yet they could also see good health, vigor, and happiness—a people who seemed always ready to break into song and dance. The better Europeans understood Aboriginal culture, the more inclined they were to acknowledge its power to rise above the bare needs of livelihood. In the richer parts of Arnhem Land, only three days of work in a week can satisfy the requirements of subsistence. Plenty of time is left over for other things. The intricate web of trade routes crisscrossing the continent is evidence of surplus. A large volume of nonutilitarian articles, most of them products of leisure, passes through this web.[5]

In areas rich in art, most men can paint, carve, and incise. Some, however, are recognized as better than others. The skill of certain songmen and dancers is so greatly valued that their services are in almost constant demand. Tradition largely dictates form and content; nevertheless, within it individual talent can and does flourish with the encouragement of the group, all of whose members are to varying degrees practicing artists themselves. A talented person who also has a strong personality may attract disciples. His method improves with practice, and his range of themes steadily expands. In time, he establishes a "school." Regional variations in Aboriginal art are thus consequences not only of geographic separation and of adjustment to local resources, but also of individual inspiration. "A Songman sees, or dreams, something. Words, tune and actions 'come' to him. He works it over in his mind and then quietly hums it. A new song is born. . . ."[6]

ETHOS AND AESTHETICS

"Art for art's sake" is alien to Australian Aboriginal thought. Yet in some cases their aesthetic drive overrides practical considerations. In some parts of northern Australia the utility of spears is sacrificed to beauty and form: barbs are cut so deep into the stems that they would snap if thrown.[7] Designs painted on bark, on sacred objects such as the bull-roarer and the *churinga*, and on the human body are so pleasing in composition, color, clearness of line, sense of movement, and vitality that, to the outside observer, they demand the word *aesthetic*. The artistic bent of Australian Aborigines is also attested by the readiness with which they express their experiences and worlds in alien media and alien styles, and by the speed with which they achieve excellence in them. Central Australian artists, in one well-known case, have quickly adopted crayon-and-paper, acrylic, oil, and watercolor in creating traditional designs and motifs; and in the Alice Springs region they have achieved wide recognition for their Western-style (perspectival) landscapes. These landscapes could have been painted by any accomplished Western artist except for the fact that they are not merely picturesque views, but also Dreaming sites, of deep religious significance to the local people.[8]

The aesthetic impulse, so inextricably a part of the cultural impulse toward form and meaning, is a human universal. Even when an Australian Aborigine's decision to paint or engrave a bull-roarer in a certain way is motivated by religion, the precise manner of execution—the deft touch, the care or decisiveness with which certain lines are drawn, certain colors applied—is governed by aesthetic, and not only mythic, appropriateness. Moreover, religion itself may be seen as a complex aesthetic-moral work, directed toward achieving an overarching sense of order and meaning.

Still, the differences are obvious. The world of the Australian Aborigines is not that of the traditional Chinese, if only because the Aborigines have no word for art and the Chinese do. Medieval Europeans also lacked a separate concept of art. Their world, like that of the Australian Aborigines, was suffused with religious significance. Yet in all other respects the two worlds differ radically, and all three of them have little in common with modern secular American conceptions of place and scene.

In exploring the more distinctive aspects of the Australian Aboriginal ethos and aesthetic practice, we encounter one practice that must seem strange to most people today: the hidden location of the Aboriginal artworks. In west Arnhem Land, many of the designs and naturalistic figures are painted in places that are accessible only with the help of a notched pole or ladder,[9] and the artists must have used some kind of raised platform to reach the cave roof. This and the curious fact that the drawings often overlie one another have led A. P. Elkin to conclude that "satisfaction lies not so much in admiring the finished picture, as in the act of painting it or in some practical desire it expresses and in some result it will effect."[10] Another distinctive trait of Aboriginal art is the importance of the multimodal performance. It may be that the prehistoric roof designs and figures were themselves part of a larger artistic-ceremonial enterprise. We cannot know for sure. We do know that contemporary rituals of any scope require the service of not just one or two but several aesthetic skills. In their totemic and initiation rites and in their multitribal celebrations (*corroborees*) the Aborigines sing, chant, dance, play act, or mimic; and the words in a song can have all the marks of the close observation and feeling of poetry.

Fresh water running, splashing, swirling,
Running over slippery stones, clear water,
Carrying leaves and bushes before it,
Swirling around. . . .[11]

SONGLINES AND LANDMARKS

The unique feeling-tone of the Aboriginal world derives from the centrality of its belief in an age of Dreaming. During dreamtime, powerful beings—ancestors and heroes—walk the earth, establish the topographic features, call all the natural species to life, and institute rules of group and individual behavior. "Walking the earth" has a literal meaning. The ancestors emerge somewhere, but rarely stay put. In a few instances, their first task is to introduce some sort of cosmic order to inchoate reality. Thus, according to one myth, "two Men brought light to the world by pushing the sky away from the earth with their sacred boards." Most myths, however, tell of incidents in the lives of dreamtime ancestors as they travel, each of which leaves a mark on the land. Natural features such as prominent outcrops, hollows, waterholes, and old tree stumps are places where a notable event has occurred: thus, a kangaroo or snake, killed by an ancestor, turns into a rock; spears left behind turn into clumps of trees; blood from a wound turns into a spring of fresh water.[12] Sometimes the ancestors call landscape features into existence by naming and chanting. Creation occurs by means of a song; and since the stories of ancestral walkabouts are now told in songs and since the myths are ritually enacted to the accompaniment of much singing, it is as though the Australian landscape is a musical map or score, and the tracks of the ancestors what Bruce Chatwin calls "Songlines."[13]

At a practical level, the dream tracks or Songlines are an effective memory aid, enabling the Aborigines to familiarize themselves with the essential landmarks and resources of a strip of country through which they must travel in search of food. At an emotional level, the animals, tracks, and other features of the landscape are all powerful reminders to the Aborigines of their own past—of how things and social practices have come into being and what they mean. This past merges with the present. Although ancestors are often depicted as having died, this death is inconclusive, for they may still reveal themselves in the power of water, blood, and fire, or as rainbow and marks on a rock, or as the bull-roarer, on which the incised or painted signs project the present power of the dreamtime spirits. Time is also conflated in the cult ceremonies, which are either social or totemic. Social ceremonies are essentially instructional. Singer-actors reenact the dreamtime of ancestors and heroes; but although the didactic purpose is made clear by the pauses in the performance, during which an old man explains its meaning to novices, the singer-actors may be wholly engaged in the performance itself. They enter into semitrance, draw blood from themselves, merge the present with the past, who they are with the roles they have assumed. Totemic ceremonies are enacted to promote the increase of species. Participants undergo the same sort of events as the original heroes and assume their generative actions, often at the sites sanctified by them. In a totemic ceremony, participants are not merely performing historical roles but have, in a sense, *become* the mighty beings of dreamtime.[14]

The earth, to the Aborigines, is thus rarely "landscape," as that term is understood in China and the Western world. It is not a view or prospect; rather, it is a particular site, a succession of sites along a strip of land, a dream track or a Songline, vibrant with incident, power, and meaning. The power and the meaning are out there in the features of the land. Rocks and waterholes are live presences. Aborigines would probably not call them "beautiful," in its restricted modern sense. And yet, when they recount in the poetry of songs and chants how these features came into being, and when they express their feelings about them in the rhythms of dance and in the composition and color of visual art, it is clear that the Aborigines' perception of and engagement with their world is both dramatic and aesthetic.

Nature and Landscape in China

The words *dramatic*, with its sense of action and emotional engagement, and *aesthetic*, with its implication of distance and reflection, are also applicable to the bond between the Chinese and their natural world. The precise character of that bond, however, differs from that of other peoples. Thus, unlike the Australian Aboriginal story, which lacks a clear sense of progressive development, the long and well-documented Chinese story reveals a general trend from an emotional and all-involving period to one that was comparatively cool and reflective.

Confronted by the strange and the unknown, the Chinese, like most humans in similar circumstances, initially felt fear. To a poet who wrote in the Earlier Han dynasty (202 B.C.–9 A.D.), South China's forests, with their "twisting and snaking" trees, their tigers and leopards, held terror rather than beauty.[15] In time, fear modulated into awe and reverence, and nature, though still regarded as a great power, was viewed as neutral or even benign rather than malignant and arbitrary. A sense of underlying harmony prevailed. Mountains, rivers, forests, and other natural features were all considered to be the local embodiments of cosmic energy, which waxed and waned so that at any time one force might be dominant, and yet in the longer run they balanced each other. The common people tended to view certain natural landmarks not as abstract forces but as deities and habitations of deities and spirits, who, together with the spirits of human ancestors, lived in accord with one another and with humans. Reciprocity was the governing moral-aesthetic principle. The spirits bestowed benefits such as male children and good harvest; in return, they expected sacrificial offerings and respect.[16]

Fear tempered first to awe and finally to appreciation. In place of a dark and ominous force to be avoided or propitiated, in place even of a sacred but still dangerous presence to be approached with circumspection, a harmonious universe emerged to Chinese eyes—grand and spectacular, fragile yet eternal—that satisfied their deepest spiritual longings. This was landscape.

SHAN SHUI

Landscape—that is, "mountain and water" (*shan shui*)—is a more domesticated version of heaven and earth (*t'ien ti*). A cosmic aura still attached to landscape, as did awe and intimations of the sacred, and indeed vestigial beliefs in fertility and potency; but by the third century A.D. landscape became essentially a beautiful-sublime concept and a world that the viewer would want to and could enter. Increasingly, the refined aesthetic sentiments of Taoism and Buddhism covered up the rawer religious emotions.[17]

The idea of landscape is manifest in garden design, poetry, and painting. Some of the earliest examples of the landscape ideal come from descriptions of the great imperial hunting parks of the Earlier Han dynasty. One such park, built outside the capital city of Ch'ang-an, was immense, surrounded by a circular wall of more than "400 *li*." Within it were mountains, thickets, forests, and marshes; in addition, there were artificial lakes and islands, plants and animals from distant countries, and thirty-six palaces and buildings. Taoist magical beliefs inspired the artificial components;

thus pyramidal islands rose out of man-made lakes in imitation of the three legendary Blessed Isles in the Eastern Sea. The entire park was a rather confused microcosm, created to satisfy the needs of both profane activity and immortal longings. In it the Emperor Wu hunted lustily. After the slaughter, he and his entourage feasted and were entertained by dancers, clowns, and jugglers. At the end of these noisy festivities, he would climb one of the great towers that commanded sweeping views, and there commune with nature in solitude.[18]

When Han poets described a landscape, they were more likely to describe a great hunting park than truly wild mountains and rivers, of which they had little personal experience. When Han artists painted landscapes on palace walls, these were more likely to show the anthropomorphized spirits of mountains and rivers than the natural features themselves. Sensitivity to nature, free of references to magic and spirits, began to emerge after the breakdown of the Han dynasty and during the later part of the Period of Disunion (ca. 200–600). Consider a poem by the poet-recluse T'ao Yuan-ming (365–427). It evokes a typical Chinese landscape, and one can imagine its appearing as a colophon on a landscape painting. The poem speaks of the poet himself in the seclusion of his country home. He putters about in his garden. He leans on his staff and sometimes lifts his head and looks around. He sees "clouds idly climbing the valleys and birds, weary of flying, seeking for their nests." Light dims, but the poet-recluse remains in the fields, "caressing with his hands a solitary pine."[19] The sentiment for landscape clearly existed in T'ao Yuan-ming's time, but paintings of such sentiment and landscape did not yet exist. Visual artists in the Period of Disunion lacked the skill to depict on paper and silk a panorama of mountains and valleys wrapped in clouds and mist. That skill was to come later.

EREMITISM AND THE CULT OF FEELING

A nature aesthetic is a refined product of civilization. Its continuing existence and development depend on the presence of civilization as antithesis. In China, withdrawal to a secluded place in the midst of nature first gained popularity when the vast and sophisticated Han empire was in turmoil; and although the power of the ideal fluctuated throughout traditional China's long history, it never ceased to exercise some appeal to the intelligentsia. Mandarin gentry who were the empire's scholar-officials fulfilled their ambitions at court, but they also found life there constraining, vexing, and sometimes dangerous. Many were torn, at least occasionally, between the splendor of the city and the Taoist-Buddhist allure of the countryside. Some disdained official life altogether and aspired solely to the virtue and charmed life of the artist-recluse.

Chinese poetry and painting give the initial impression of a people with a deep love for nature and landscape. However, as one becomes more familiar with the history and philosophy of these works, the question arises as to whether that love was genuine. Had the literati-artists who produced the poems and the paintings any firsthand knowledge of wilderness or, for that matter, the real countryside? Tradition exerted a strong grip on Chinese society, and nowhere was this more evident than in the practice of art. The imitation of the style of illustrious predecessors was not so much evidence of a second-rate talent as of a sophisticated knowledge of the sources of excellence. A tendency in certain periods to turn inward to feeling further militated against direct experience and observation. An influential figure of the Sung dynasty, Su Tung-p'o (1037–1101), argued that the purpose of painting was to express the artist's own feeling rather than to depict the external world. He and his friends, all gentleman-scholars, maintained that the mere accurate portrayal of things could best be left to professionals.[20] Again, in the Ming dynasty (1368–1644), prominent artists urged that meaning was to be found "within" rather than "out there." Shen Chou explained in an inscription how he came to paint Night Vigil in 1492. He woke up in the night, his mind clear and untroubled. "How great is the strength to be gained sitting in the night. Thus, cleansing the mind, waiting alone through the long watches by the light of candle becomes the basis of an inner peace and of an understanding of things."[21] With such understanding already in the grasp of the artist, what need was there to step out into the night and observe?

Eremitism implies solitude and withdrawal from society. However, increasingly, the gentry found it more convenient to withdraw figuratively rather than literally. The attitude of Wen Chen-ming (1585–1645) was fairly typical: "To live in the far country is best, next best is to live in the rural areas, next comes the suburbs. Even if we are unable to dwell among cliffs and valleys and follow the path of the hermits of old, and have to settle in city houses, we must ensure that . . . the rooms are clean and smart, that the pavilions suggest the outlook of a man without worldly cares, and that the studies exude the aura of a refined recluse."[22] From the Yuan dynasty onward, the affluent favored the suburb rather than either the countryside, considered too remote and lacking in amenities, or the city, considered too crowded and hectic. By the late Ming dynasty, however, the city itself attracted an increasing number of people who, although they obtained their income from commerce, also had literary-artistic aspirations. For them, the solution to having the best of both worlds—closeness to the urban amenities and the enjoyment of solitude—was the garden.

THE GARDEN

Imperial parks of the Han dynasty were conceived on a huge scale to suggest Mount K'un-lun to the west, the Blessed Isles of the Eastern Sea, and the haunts of nature deities and immortals. In contrast, city gardens of the gentry in the Ming dynasty were small, many less than one acre in size. Yet they also had cosmic pretensions. A garden was artfully designed to give visitors the impression that they could wander through its intricate paths, among hollows and craggy "mountains," and pause over arched bridges and in secluded pavilions for days on end without exhausting its wealth of prospects.[23] To overcome the actuality that this space was bounded by the blank face of a wall, imagination must come powerfully into play.

In certain lights—at dawn or dusk, or sometimes in the blinding glare of summer noon—the wall might even seem to have melted away altogether, leaving the rocks and bamboo floating in . . . horizonless and vaporous distances. . . . Shrinking himself in imagination to the size of an ant, the connoisseur could wander in these misty wastes among rocks now grown into mountains, and shrubs and grasses as big as trees and forests. And as he walked and paused, the landscape unfolded around him as if he were taking a three-dimensional stroll through one of his own paintings, slowly unrolling the horizontal scroll from right to left.[24]

FIELD OBSERVATION AND REALISM

Not all gentry viewed nature as pure artifice. Some prominent artist-poets practiced the *yang* of direct field observation, as opposed to the *yin* of feeling and inwardness. Beauty was "out there," and one must go out to meet it. The landscape poet Hsieh Ling-yun (385–433) had a special pair of climbing boots made so that he could gather herbs in the hills and experience the sheer joy of climbing.[25] The Ming artist Li Jih-hua recounted that the Yuan master Huang Kung-wang (1269–1354) "used to spend whole days sitting amidst the deserted mountains, disorderly rocks, thick forests, deep water-falls, feeling utterly free. People could not understand why." But it was as a result of this activity, Li explained, that Huang's "brush can be so spirited and transforming that it competes with the wonders of nature."[26] "Take Heaven and Earth as models rather than the old masters," urged Tung Chi-ch'ang (1555–1636). "One should ob-serve every morning the changing effects of the clouds, break off from practicing after painted mountains and go for a stroll among the real mountains."[27]

Westerners exposed to Chinese landscape painting for the first time may be struck by its ethereal beauty but wonder whether this world of sheer cliffs and mist is real. A visit to China will show that it is. Unlike much of Europe and the midwestern United States, where rolling ridges and valleys are the norm, China's topography is dominated by mountains and hills that rise sharply out of flat allu-vial plains. When autumnal mist rolls in, not only Hua Shan in north China and Huang Shan in south-central China (famous quasi-sacred mountains well known to artists), but even the modest hills and plains of An-hui, Che-ch'iang, and Ch'iang-su—a center of buzzing artistic activity during the Sung and Ming dynasties—take on the otherworldly appearance of Chinese landscape paintings.[28]

Fidelity to nature was a desideratum of artists in the tenth century. It was said of T'ung Yuan, who died in 962, that he "did not paint strange peaks"—a claim that can mean either that he avoided the "impossible" cliffs and chasms of the north or that he did not follow the fantastic topographic distortions of the late T'ang expressionists. Artifacts were also closely observed. Kuo Chung-shu, who died in 977, showed a precise knowledge of how river boats were built in his *Traveling Upriver in Midwinter*. In the eleventh century (Northern Sung dynasty), artists tried even harder to present the external world accurately. A last great monument of Northern Sung realism, according to Michael Sullivan, "is a handscroll painted between 1100 and 1125, which depicts life along the river outside the capital at the time of the Ch'ing-ming festival in early summer. The painter [Chan-tse-tuan] takes us on a leisurely walk by the riverside, over the bridge, through the city gate, and into the busy streets, as though he were tracking with a movie camera." All the techniques at his command, which included shifting perspective, foreshortening, and shading, were used to produce the impression of a real world.[29]

Some Ming artists turned inward and to exemplars of the past for inspiration. Others sought inspiration from both interior and exterior sources. Shen Chou (1427–1509), for example, favored withdrawal into a room to seek an "understanding of things" by the light of a candle; but he was also a dedicated traveler and painted the places he personally knew. He wrote: "I know the route [around Suchou] intimately and not a year goes by without a visit to familiar spots. Many are those, which in response to the eye, I have committed to the brush as painting and poems."[30] His pictures capture some of the more homely aspects of landscape. An example is the scene around the canal city of Kao-yu. It shows "the inside curve of a city wall, the partially opened gate; outside, temple pagoda and roof; then flat openness—six water-bordering willows, a boat with yet-to-be-raised sail accenting the arteries of travel, tiny farmers irrigating their rice-fields, boundary paths of those fields. Two ploughmen and their water-buffalo. Distant roofs beyond the fields. . . ."[31]

The style of Shih-t'ao, a Ch'an Buddhist priest of the early Ch'ing dynasty (1644–1912), is often abstract and free, notably in a painting called *Ten Thousand Ugly Ink Dots*. Yet even there it is not difficult to recognize a rather typical Chinese landscape. In any case his realist credentials are firmly established by his portrayals of one of China's legendary peaks—the Huang Shan in An-hui Province. He visited it first in 1667, again in 1669, and during a month's stay painted seventy-two views. His *Album of Landscapes* (1701) contains pictures of other "strange peaks" of southern An-hui, and despite the fact that they were done later from memory in Yang-chou, they clearly show respect for the region's topography.[32]

ENDURING VALUES AND BELIEFS

The relation of the Chinese gentry to nature is complex, subtle, shifting, and at times contradictory. Artists and scholar-officials appear to be torn by opposite values, desiring both nature's quietude and the city's stimulation, both knowledge that comes from direct contact with nature and knowledge that comes from withdrawal and introspection, both wild mountains and the cultivated human world. Nevertheless, the evidence of landscape art suggests that certain enduring values and beliefs exist. One such value, not obvious from a superficial look at landscape art, is livelihood; some ordinary human activity nearly always appears in a landscape. Fishermen fishing from their boats and travelers (merchants and pilgrims) negotiating a mountain path are among the most popular. In Shen-chou's *Kao-yu*, we see not just a human figure or two, but a countryside of ricefields, plowmen, and water buffalo. Livelihood brings fertility to mind: animals and plants must multiply for humans to multiply and prosper. Nature as generative power is an enduring concept. Sexual symbolism is evident in pictures of towering peaks and of rivers emerging from chasms. Nature is also regarded as a force that affects the human world. Hsia-kuei's *Sailboat in the Rain* (early thirteenth century) shows a summer storm bending trees and swelling the sail of a boat. His *Winter Evening* reveals a nature that closes in on the human world, limiting people's vision and constraining travel.[33] Nature and human emotions fuse. Artists use nature to express emotion. "I have painted orchids with an air of happiness, and bamboos with an air of anger," wrote an artist-monk of the Yuan dynasty (1260–1368).[34] Such practice leads to the belief that nature *has* these feelings and hence is kin to man. Indeed, nature may go out of its way to respond to human wishes and desires. For example, when Shih-t'ao went in search of plum blossoms near Nan-ching, he felt as though the mountain there was bending toward him, the more readily to display its wonders. The artist captured the experience in a remarkable painting called *Mountain Bends to Man*.[35]

The principal source of inspiration for Chinese landscape painting is Taoism-Buddhism's fundamental mystical outlook on reality. This outlook cannot be conveyed fully in words; hence the value of the picture, in which one typically finds in the foreground ordinary objects and events and, in the background, a world that, while it is recognizably *this* world, also strongly hints of another. In the foreground is, say, a hermit contemplating a waterfall or a fisherman putting up sails; also in the foreground are such natural features as an oddly shaped rock, a river bank, a lotus blossom, a tuft of bamboos. Beyond them, one is led outward by gradual steps through the partially overlapping features of the middle ground to the boundless universe.

The Medieval
European Cosmos

Appreciation of landscape is an urban aesthetic: its efflorescence depends on the existence of a cultivated elite who live, at least part of the time, in large, cosmopolitan cities. Medieval Europe lacked such cities. Mountains and forests, where these pressed close to human settlements, were regarded as threatening. The idea that one might stand on a prominence and enjoy a view did not arise until late in the Middle Ages except in the case of unusual individuals such as the poet Petrarch, who was known to have strolled in the mountains under moonlight.[36]

NATURE SEEN CLOSE AND SYMBOLICALLY

Rather than views and prospects, people in the Middle Ages enjoyed nature close at hand; and, at quite another level, they rejoiced in heaven, where they expected to go after death. The middle ground and the middle scale were relatively unimportant. "Nature, for Chaucer," notes C. S. Lewis, "is all foreground; we never get a landscape."[37] Love of small objects in the foreground can be touching, as the following lines from Chaucer's *The Legend of Good Women* show:

> *And now on knees aright I me set,*
> *And as I could this freshe flow'r I grette,*
> *Kneeling always till it unclosed was*
> *Upon the small, and soft, and sweete gras.* (115–118)

Increasing attentiveness to nature's intricate forms and patterns is evident in the decorations of Notre-Dame de Paris. Botanical motifs in the choir, completed about 1170, were still abstract and geometrical. "Ten years later, when the first spans of the nave were built, the flora that adorned them was already closer to its living models; deliberate symmetry had vanished and the diversity of actual nature was visible, so that it is possible to identify a given leaf or distinguish a given species. Yet even so these plants were still chiefly symbolic, and it was not until other parts of the edifice were decorated, after 1220, that they became true to life."[38] Works produced with such care suggest that their creators derived sensual-aesthetic pleasure from the shape of a leaf or a tendril, the plenitude of a cluster of grapes. But even in the thirteenth century, it would be inaccurate to say that the medievals liked nature for itself alone. God still made his presence felt everywhere. Everything in nature revealed his visage, the daisy no less than the sun.

Saint Francis most famously loved nature both in itself and sacramentally. "When he considered the glory of the flowers, how happy he was to gaze at the beauty of their forms and to enjoy their marvellous fragrance," wrote the saint's first biographer, Thomas of Celano. But the experience did not stop there; the saint's spirit rose to "meditate on the beauty of that unique flower," his Lord and God.[39] By the fourteenth century, religious thinkers deemed all created things worthy of love and, indeed, in a figurative if not literal sense, capable of offering praises to their creator. The Dominican monk Heinrich Suso addressed God with a lyricism echoing that of Saint Francis's *Canticle of the Creatures*:

> O admirable Lord, I am not worthy to praise thee, yet my soul desires that the sky may praise thee when, in its most ravishing beauty, it is illuminated in full clarity by the brilliance of the sun and the countless multitude of scintillating stars. That the fine countryside may praise thee when, amid summer delights, it sparkles in its natural dignity, in the manifold array of its flowers and their exquisite splendor.[40]

THE HEAVENLY VAULT

Suso's song of praise enlists, on the one hand, flowers in the summery countryside and, on the other hand, the sky and its stars. Flowers are nature close at hand. Their exquisiteness is easily appreciated. The overarching sky is distant. Its beauty is accessible to the eyes, but the true dimension of its grandeur—for the medieval poet and religious thinker—lay in a conception rather than in what uninformed eyes could see. Medieval ideas about the upper world gave it a radiance that we moderns, lacking such ideas, can scarcely imagine.

The premodern European cosmos had an absolute Up and Down. Earth lay at the bottom. Above it soared the successive spheres of heaven to a great height. Looking up, one might feel awe but not bewilderment, for heaven, like the interior of a great cathedral, was bounded space.[41] The infinite and eternally silent space, which frightened Blaise Pascal in the seventeenth century, was yet to come. The space of the medievals, though vast, was not infinite, and it was filled with the eternal hymning of the angels and the music of the spheres. This globe of sound found a visual analogue in the vaulting harmonic spaces of Gothic architecture. The cathedral, though built of stone, was an embodiment of motion in music, and also of cyclical manifestations of God's work such as "the regular movements of time and the seasons . . . , the rhythms of nature, [and] the motions and humours of biological life."[42]

Outer space, to us, is not only silent and unbounded, but pitch black and deathly cold. The medievals held a very different view. Their sun illuminated and warmed the whole universe, and night was merely the narrow conical shadow cast by the earth (Dante, *Paradiso* IX. 118). In the region beyond the crystal sphere of the outermost star lay "the very heaven and full of God." When Dante passes that farthest sphere he is told, "We have got outside the largest corporeal thing into that Heaven which is pure light, intellectual light, full of love" (*Paradiso* XXX. 38).[43]

Deeply held beliefs affect how one sees. A medieval poet, looking up at the cerulean or night sky, could not help but be swayed by the ideas prevailing in his own time. But evidence directly accessible to the senses must also exist: stars have to be seen to sparkle to sustain the belief that they are luminous intellectual beings, and one needs to have experienced a perfect effulgent day filled with sunlight to think that, yes, heaven may be like that, only more so. In the Middle Ages, however, it was not nature's vault, which could be overcast and gloomy, but rather a preeminent achievement of culture, the cathedral, that provided the most tangible hint of the glory to come.

THE CATHEDRAL AND THE MYSTIQUE OF LIGHT

Unlike the sanctuaries of most ancient religions, the medieval cathedral was open to all sorts of people—prince, bishop, craftsman, peasant, the old and the young; and even to such animals as dogs and sparrow hawks that accompanied their human masters. On ordinary days people moved freely in and out of the consecrated space, gossiping and socializing. On feast days they came from all parts of the city and beyond to worship God and adore his saints. For the vast majority of people, daily life was drab and often harsh. Even the humblest, however, had access to the cathedral—"the forecourt of heaven." They had the right to be there even only to rest and sleep; and, of course, they were enjoined to be there as worshipers.[44]

The root inspiration for the cathedral was the Celestial Jerusalem of the book of Revelation rather than the gospels, which depicted lives too close to the everyday existence of the people to satisfy their yearning for a brilliant world far above the grime and poverty they knew so well. The Celestial City was a jewel of jewels, wrapped in light and color. "Its light was like unto a stone most precious, even like a jasper stone, clear as crystal. . . . And the city had no need of the sun, neither of the moon, to shine in it: for the glory of God did lighten it, and the Lamb is the light thereof" (Revelation 21:11,23).[45]

Medieval people of all ranks loved light and bright colors.[46] Umberto Eco writes: "Figurative art confined itself to simple and primary colours. . . . It depended on a reciprocal coupling of hues that generated its own brilliance, and not on the devices of chiaroscuro, where the hue is determined by light. In poetry, too, colours were always decisive, unequivocal: grass was green, blood was red, milk snowy white." The love of sparkle and bright hues was evident not only in art but in everyday ornaments, clothing, and even weapons. "Things are called beautiful when they are brightly colored," wrote Saint Thomas Aquinas in his *Summa Theologiae.*[47] Presented with light and color, medieval people were filled with the joy of life. God was perfection and pure light; the angels and archangels were luminous beings, and the faces of saints glowed with an inner light. Even gross matter, since it was God's creation, must contain light, however dim. Metals, after all, shine when they are polished. And, Saint Buenaventure asked, are not clear window panes manufactured from sand and ashes, and is not fire struck from black coal, and is not this luminous quality of things evidence of the existence of light in them?[48] Among material substances, jewels occupied a privileged position. Medieval jewels were not faceted, as was the custom during the Renaissance, but were rounded and polished so as to give an impression of light glowing from within; indeed, some people thought that light did glow from within.[49] Because of this property, jewels were credited with Christian virtues. The display of precious stones in vestments, on vessels, and in churches was thus not mere worldly ostentation.[50]

The rose window of a Gothic cathedral is a vast jewel of many colors. Sunlight, rather than sending its rays into the interior, is held back within the glass. Medieval glassmakers knew how to manufacture clear and transparent glass, but in the view of the scholar J. R. Johnston, they made a conscious effort to enrich the medium so the rays that reached it could be "scattered again and again in a complicated implosion, producing the variety and richness normally associated with old glass" and giving it that radiant interior quality.[51]

Suger (1081?–1151), the abbot of Saint-Denis and "creator" of the Gothic cathedral, was, says Erwin Panofsky, "frankly in love with splendor and beauty in every conceivable form."[52] It is hard for us to say now to what extent his attitude to gold, precious stones, stained-glass windows, and the illuminated interior was aesthetic rather than religious. We may deem it merely aesthetic, tainted perhaps by a profane love of luxury. Suger himself no doubt thought otherwise. Here is an example of how he viewed the treasures at Saint-Denis:

> Often we contemplate, out of sheer affection for the church our mother, these different ornaments both old and new; and when we behold how that wonderful cross of St. Eloy—together with the smaller ones—and that incomparable ornament commonly called the "Crest" are placed upon the golden altar, then I say, sighing deeply in my heart: "Every precious stone was thy covering, the sardius, the topaz, and the jasper, the chrysolite, and the onyx, and the beryl, the sapphire, and the carbuncle, and the emerald."[53]

The beauty of the house of God "calls me away from external cares," Suger wrote. "I see myself dwelling, as it were, in some strange region of the universe which exists neither entirely in the slime of the earth nor entirely in the purity of Heaven; and . . . by the grace of God, I can be transported from this inferior to that higher world in an anagogical manner."[54] Suger, like other religious thinkers of his time, saw art and terrestrial beauty as foreshadowing the heavens that would open up to souls after Judgment Day. "The golden door foretells to you what shines here within"—within the abbey, but also within the hearts of humans and the heart of the world itself. "Through palpable, visible beauty, the soul is elevated to that which is truly beautiful, and rising from the earth, where it was submerged, an inert thing, it is resuscitated in heaven by the radiance of its glory."[55]

Nothing more directly prefigured God and his abode in heaven than light. The patron saint of France, Saint Denis, after whom the abbey was named, was mistakenly identified with a Syrian monk of the sixth century, whose major work, *On Celestial Hierarchy*, had a powerful influence on Suger. Early on, the work asserts that "material lights" mirror the "intelligible ones" and, ultimately, the true light of the godhead itself. "Every creature, visible and invisible, is a light brought into being by the Father of lights." According to Panofsky, "The decisive feature of the new [Gothic] style is not the cross-ribbed vault, the pointed arch, or the flying buttress";[56] nor is it the soaring height, an ideal already attained in Romanesque architecture. It is light. In the words of Otto von Simson, "The Gothic wall seems to be porous: light filters through it, permeating it, merging with it, transfiguring it. The stained-glass windows of the Gothic are structurally and aesthetically not openings in the wall to admit light, but transparent walls. The Gothic may be described as transparent, diaphanous architecture . . . a continuous sphere of light."[57] Suger, as he replaced the opaque Carolingian apse of his church with a new transparent choir, rhapsodized over brightness:

> *For bright is that which is brightly*
> *coupled with the bright,*
> *And bright is the noble edifice*
> *which is pervaded by the new*
> *light.*[58]

When medieval people of strong faith entered a cathedral at the moment sunlight burst through the clouds and struck the rose window, making it glow, their feelings of joy and holy calm could be encompassed by language, if at all, only if it was mystical-religious rather than narrowly aesthetic. Medieval believers expected to find, upon dying and entering heaven, a reality with which they were already familiar. What they had known tentatively on earth they would know fully in heaven as a place of transcendent beauty, filled with truth and goodness. This belief in the unity of beauty, truth, and goodness is one that we moderns have largely lost, and one that we may look back upon with nostalgia and envy.

American Place and Scene

*T*he American appreciation of environment, like that of any other complex society spread throughout a vast land, varies both regionally and temporally. To explore American values at roughly the same level of generality as those of medieval Europe and imperial China, we shall have to look at America as a whole rather than, say, New England or Utah, and at Americans rather than, say, Irish Americans or Korean Americans.

In common with the vast majority of humankind, Americans love the small intimate world that is their home and, immediately beyond it, a rich agricultural land. Distinguishing them from other peoples is a romantic predilection for process, movement, space, and scenery and, in contrast, an unusual fondness for the classical ideals of efficiency and order. The American aesthetic is tinged with religious aspiration, as in old Europe, and with moralism, as in China. The American aesthetic, too, is affected by social meliorism: the good and beautiful place is one that shows signs of progress and is capable of further progress. The future exerts a great lure. And there is yet another important difference: in sharp contrast to the austere and "elevated" conceptions of beauty embraced by the socially established in both the Old World and the New, the American aesthetic appears at times to be driven by a sense of fun. It happily accommodates what might be called a democratic and folksy fondness for the extreme, the eye-catching, the amusing, and the grotesque.

Hometown and
Agricultural Land

Reflecting on the hometown of her childhood, Xenia, Ohio, Helen Santmyer writes: "Children are not aesthetically blind; we could not have thought that view [from the viaduct to the familiar houses] beautiful. Yet I am convinced that we must have resented any criticism of it. It was home: the same, day by day, year after year—eternally, we should have liked to believe."[1] One's home or hometown need not be beautiful by artbook standards. Other values may be more important, such as comfort and security, a haven of human warmth. Yet even in the most humdrum town there are moments of beauty. "You passed the doctor's office, and were at the corner of your own street, where you turned west, and saw the trees arched against a glowing sky." Suppose the skies were gray, "the pavements streaked with soot, and lumps and black snow filled the gutter," and you, as you paused at the gate, were wishing to escape from it all to something different and brilliant far away. Even then, at that same instant,

> you were aware of the iron of the gate beneath your hand, and were storing away the memory of how it felt. Thus the unfastidious heart makes up its magpie hoard, heedless of protesting in-

telligence. Valentines in a drugstore window, the smell of roasting coffee, sawdust on the butcher's floor—there comes a time in middle age when even the critical mind is almost ready to admit that these are as good to have known and remembered, associated as they are with friendliness between man and man, between man and child, as fair streets and singing towns and classic arcades.[2]

Santmyer, like most people, associates the aesthetic almost exclusively with the visual—with the architecture of "fair streets and classic arcades." But the texture and hardness of the iron rail and the fragrance of coffee are also elements of our total aesthetic experience. Indeed, home and hometown are loved and appreciated more through the "warm" proximate senses than through the "cool" distant sense of sight. As to what one *sees* in small-town America, true, not much of it is Art, although local citizens may take pride in "the green tiles on city hall, the Greek pillars of the bank, and the lilacs." The truly beautiful may escape the conventionally educated eye. It is, as a youngster says of his hometown in west central Texas, what happens after the sun has gone down, and the vapor lights on the tall aluminum poles over the highway start to shine. "Everything on earth is sort of gray by then, yes, lilac gray, and there are shadows down the streets, but there, while the sky is changing, those lights are the most beautiful things in the United States!"[3]

To the farmer, rich agricultural land—the "furrows rimmed with gold" to which Homer sang paeans millennia ago—has the greatest appeal. American husbandmen were no exception. Land unfit for agriculture, especially treeless mountains, was considered simply bad by early settlers. "Their contempt," writes John Stilgoe, "endured [into the nineteenth century], probably warping national aesthetics and divorcing American notions of landscape beauty from the standards dear to European romantics." Of upstate New York, Timothy Dwight wrote in 1804, "the phrase *beautiful country* . . . means appropriately and almost only lands suited to the purposes of husbandry, and has scarcely a remote reference to beauty of landscape." According to Stilgoe, even the "wilderness portrayed in so many paintings and tales reflects the aesthetic of the agricultural land concerns." Wilderness is beautiful if it is either symbolic of nature's plenitude or promises future productivity. Only when explorers entered the High Plains and the Rocky Mountains, after the 1870s, did rocky and barren scenes—a mineralogical topos—enter the American consciousness as in their own way beautiful.[4]

Romantic
Wilderness Prospect

Wilderness prospects, provided they do not connote infertility, have enchanted well-educated Americans since the eighteenth century. As early as 1728, William Byrd of Virginia proclaimed delight in the Appalachian Mountains, in the "Ranges of Blue Clouds rising one above another." Once, when fog prevented a clear view, Byrd lamented "the loss of this wild prospect"; its lifting, however, rewarded him by dramatically opening "this Romantick Scene to us all at once."[5] Thomas Jefferson's description of the view at the Blue Ridge water gap is rightly famous for its vividness, but also for its success in capturing several key qualities that together make the prospect enthralling: the vastness of geologic time and of geographic space, tumult and serenity, the lure of distance, and a "fine country" not far from "terrible precipices."

> For the mountain being cloven asunder, she presents to your eye, through the cleft, a small catch of blue horizon, at an infinite distance in the plain country, inviting you, as it were, from the riot and tumult roaring around, to pass through the breach and participate in the calm below. Here the eye ultimately composes itself; and that way, too, the road happens actually to lead. You cross the Potomac above the junction, pass along its side through the base of the mountain for three miles, its terrible precipices hanging in fragments over you, and within about twenty miles reach Fredericktown, and the fine country around that. This scene is worth a voyage across the Atlantic.[6]

Romanticism was a powerful strand in European sensibility at a time when Americans grew conscious of their own nationhood and of the vast land on which it was to be built. Americans took naturally to romantic scenes and panoramas. By the nineteenth century, Europeans already had to travel some distance to experience wild nature. Americans could still find it at their doorstep. Yet, in their imagination, wilderness was remote—a remoteness that lent it enchantment. Much of the approbation was highly self-conscious. A gentleman felt obliged to show his sensibility. "By the 1840s it was commonplace for literati of the major Eastern cities to make periodic excursions into the wilds, collect 'impressions,' and return to their desks to write descriptive essays which dripped with love of scenery and solitude in the grand Romantic manner."[7] But the wild prospect was far from being a mere literary and artistic cult. Even Americans without literary-artistic pretensions flocked to see, first, paintings, then photographs of panoramic scenes. Photography, introduced into this country in 1839, quickly produced acceptable landscapes. But for a long time it was no match for painting, if only because it could not capture the colors of a sunset or an impression of turbulent air and cascading water, and also because it could not, without a high pedestal on which to place the bulky camera, create a sweeping bird's-eye view—a sense of boundless space that an artist, if he had the skill and imagination of a Frederick Church, could evoke. In our time, the mythic quality of cowboy movies depends on the West's austere and sculptural landscape. Quite apart from the gunfights, a panorama that stretches across the wide screen and floods the viewer's field of vision is in itself dramatic.[8]

Classical Appeal:
Order on the Land

The eighteenth century was also the Age of Reason. Classical values of proportion and geometry enjoyed American favor alongside romantic dynamism, open vistas, and wildness. Place-names such as Athens, Rome, Troy, Syracuse, Ithaca, Utica, and Augusta dignified villages and towns. In architecture, Greco-Roman styles and motifs on state capitols, colleges, mints, courthouses, and post offices lent quick *gravitas* to an otherwise disheveled, "romantic" landscape. "Until the rise of the international style," writes Howard

Mumford Jones, "classical forms were standard for public build-ings."[9]

But the most distinctive and widespread classical element in the American landscape is not the quaint Greek or Roman place-name or the pediment of a bank facade but the rectangular system of United States land survey and the grid pattern of so many towns and cities. Jefferson drew the first plan for a survey of the Public Domain in 1784. His grid was made up of townships, each one hundred square miles in area, rather than the thirty-six-square-mile size that was adopted in 1785. The difference between the two fades into in-significance compared with their shared ideal of rationality, propor-tion, and orderly process. Social harmony itself was and is believed to be a fruit of rationality—of "order on the land." Who knows but that the rectilinearity (the rectitude, as it were) of the survey has con-tributed measurably to public peace, and that the Wild West so fondly projected into the nineteenth century would have been a great deal wilder without its defining power—its sobering lucidity.[10]

The grid pattern of towns and cities was motivated by crass ma-terialism: it facilitated the sale of property. But it also promoted or-derly process and efficiency of settlement and, moreover, projected an air of welcoming openness to strangers. A grid town is quickly known—a fact appreciated by people passing through. By the same token, it could quickly seem monotonous. American towns have often been criticized for lacking the interest and the beauty of older settlements. There is not only the monotony of the street pattern; there is also the sameness of "brick buildings along main street and freestanding frame houses, each with a lawn," along the perpendicu-lar side streets. One town is much like another. However, rather than monotony, J. B. Jackson suggests that what American towns show is conformity to a distinctive American style. "*Classical* is the word for it. . . . Rhythmic repetition (not to say occasional monotony) is a Classical trait, the consequence of devotion to clarity and order."[11]

Moralism

Most people the world over do not distinguish between "good" and "beautiful." Agricultural land is good, hence also desirable and pleasing. Americans share this view. However, when Americans appraise a landscape, they also use the word good in an explicitly moral sense. St. John de Crèvecoeur (1735–1813), for instance, gave the moral advantage to America when he compared the new country, with its egalitarianism and orientation to the future, with Italy, rooted in the snobbisms and conflicts of the past. "In Italy, all the objects of contemplation, all the reveries of a traveler, must have reference to ancient generations and to very distant periods. . . . Here, on the contrary, everything is modern, peaceful and benign. . . . Here everything would inspire the reflecting traveler with the most philanthropic ideas. . . . Here he might contemplate the very beginnings and outlines of human society."[12] Even the backwoods scenery of ramshackle farms and fallen trees could be admired because it suggested a life of independence and resourcefulness. "The American landscape was beautiful," J. B. Jackson writes, "because it reflected a social order which was free and egalitarian. Its beauty was that of a symbol which men united in venerating."[13]

As for wild nature, it is an "august TEMPLE" in which we dwell "for lofty purposes. Oh! that we may consecrate it to LIBERTY and CONCORD, and be found fit worshippers within its holy walls!" The sentiment and the style are representative of their time: 1835. Much of that feeling (though not the style) lingers into our time. One ventures into the wilds to commune with their preternatural presences and returns to workaday life a better person. In the mid-nineteenth century, although both European and American literary figures took to nature romanticism, there was an important difference. In America, according to Perry Miller, nature not only consoled and uplifted the individual; it was also credited with the power to assuage a "national anxiety," namely, descent into a totally artificial life under the pressure of civilization. Nature's immense spread in the New World provided a guarantee against that outcome. To be natural and spontaneous, Europeans were obliged to live as outcast bohemians among "the brick and mortar of Paris." By contrast, any child of the Ohio valley could flaunt both qualities without becoming a social outcast. "America, amid its forests, could not, even if it tried, lose its simplicity. Therefore let Christianity bless it."[14]

Americans not only sang paeans to farms and wilderness, they also poured encomia on the nation's spacious cities. Writers in America continued to see actual or potential wonders in the metropolis long after their confrères in Europe had become disillusioned, following the ravages done to urban life by smoke-belching industries. The idealization of the city began in New England with "The City on a Hill," and continued through the nineteenth century as the New Jerusalem, the New Rome, the Western El Dorado. Emerson spoke of "spaciousness" as though it had not only a physical but a moral meaning; Thoreau, too, when he described the American city as "well spaced all at once, clean and handsome to the eye—a city of magnificent distances." Emerson commended St. Louis for its "spacious squares, and ample room to grow"; he praised the "magnificent" hotels of Cincinnati and Philadelphia, and noted with pleasure the noble buildings and expansive vistas in Washington. Large urban space suggested to Emerson large moral ends, and he could judge a city (Philadelphia) harshly when its "miles of endless squares"—its "monstrous" size—bespoke moral vacuity rather than moral elevation.[15]

In the twentieth century, austere functional architecture, even though its origins can be traced to Europe, has always seemed quintessentially American. It belongs by natural affinity to the New World. The glass office tower in particular quickly gained favor among American architects and corporate leaders. Lithe and clean, it shoots skyward not by graduated steps or with the support of buttresses, but effortlessly; it projects vitality and efficiency. These aesthetic or quasi-aesthetic values exist alongside the moral desideratum of erecting buildings that, unlike the sham historical facades and masonry walls of the nineteenth century, which hid function and created caverns of darkness, reveal the truth of the modern world to the people who live in it—a lighter, freer, more honest world, made possible by technology.

The problem with attributing moral power to environment is that it is seldom justified: wilderness does not always console or inspire virtue, largeness of square can be *read* as largesse of spirit, but the one cannot produce the other, and the glass office tower has proven to be as opaque and impenetrable as the thickest masonry. Rather than symbolizing a more open and democratic society, glass towers, for pedestrians pounding the sidewalks, have come to stand for exclusivity, indifference to public needs, technological fantasy, corporate privilege, arrogance, and power.

Newness and
Cleanness: City Beautiful

"For now I create new heavens and a new earth, and the past will not be remembered, and will come no more to men's minds." This passage from Isaiah (65:17) must have struck a sympathetic chord among Americans, who, from the start, saw themselves as building an uncorrupted society in a New World. The Revolution reinforced this belief in beginning anew. Oldness was bad, and newness, says David Lowenthal, was "not only tolerated but positively worshipped. . . . lack of historical remains became a matter for self-congratulation."[16] Jefferson was not alone in his view that "Our Creator made the earth for the use of the living and not of the dead. . . . One generation of men cannot foreclose or burden its use to another." As Noah Webster saw it in 1825, Americans were not the descendants of the whole sorry race of mankind; rather, their "glory begins at dawn." The worship of newness applied to the material environment as well. A time will come when "no man shall build his house for posterity." So a reformer thought in Hawthorne's *House of the Seven Gables*. Emerson (Lowenthal reminds us) criticized a Massachusetts state survey for recommending stone houses, on the grounds that they lasted too long. "Our people are not stationary . . . and so houses must be built that could easily be moved or abandoned." Thoreau wanted all relics of the past destroyed. He expounded a social philosophy of "purifying destruction."[17]

Americans often note and denounce the brutal ease with which an existing urban fabric is destroyed, even when (despite dilapidation) it is still functioning, so that new housing projects, shopping malls, and skyscrapers may be erected. The usual explanation is greed—or, to put it more politely, the inexorable logic of capital investment.

Valid as such a psychoeconomic explanation may be, the aesthetic lure of the new should not be altogether discounted. New is simply better. The New Republic, after all, has constructed its identity against a backdrop of discarded Old World values. The continent itself may be old, and indeed Americans, including Jefferson, boasted about the geological age of their land as a counterweight to the lack of artifactual antiquity. More often, however, they saw nature not as old and dead (like rocks and fossils) but as a vital, self-renewing force. Nature was able to project—and still can project—an image of freshness with the power to invigorate those who venture into its midst. The healthful untrodden appeal of the wilds is contrasted with the stale calcified conventions of society.

New is clean; old is dirty. "Clean" and "dirty" have moral as well as aesthetic meaning. A pioneer farm may be untidy, but it is clean, and its dirt is clean dirt. A farm, town, or city builds on virgin soil rather than on the middens (the accumulated dirt and errors) of the past. Cleanness is a valued quality in the built environment. The musicologist Ned Rorem, visiting the White House, observes: "How unostentatiously clean [it is] with its neat unobtrusive flower arrangements, its groups of military crew cuts performing chamber music in corners!"[18] Surely one reason for the success of McDonald's worldwide is its cleanness—the crisp white paper bags that offer up hamburgers and French fries, the eager helpfulness of the servers, the trash-free inviting appearance of the place itself. McDonald's can be dismissed as sterile or merely hygienic; but the millions of customers who flock to it appear to think otherwise. The customers would not use aesthetic terms to describe their appreciation. If asked, they may say that it is "clean," "friendly," "fast," and "predictable." Yet these terms are not so distant in meaning from "sparkling," "new," "shining," "orderly," and even "harmonious," which do have an aesthetic flavor. On a larger scale, American inventions such as the shopping mall and Disneyland, which have become immensely popular since World War II, also capitalize on people's desire for cleanness—markets without refuse, frontier saloons without beer stains, life without shit.

Thoreau described the American city as clean, handsome, and

spacious. If crowding suggests dirt, open space—the broad streets that dwarf the houses on either side—suggests cleanness. Western cities and frontier towns are laid out along generous lines. However, even a city of crowded skyscrapers can project an image of cleanness if the buildings rise up like marble slabs or are sheathed in glass that sparkles under the sun. Manhattan has perhaps always been a rather dirty place. If one doesn't like it, one concentrates on the filth and the noise. But if one does like it, one might celebrate its austere verticals, in the manner of Gertrude Stein, who wrote: "I simply rejoiced in the New York streets, in the long spindling legs of the elevated, in the straight high undecorated houses, in the empty upper air and in the white surface of snow. It was such a joy to realize that the whole thing was without mystery and without complexity, that it was clean and strait and meagre and hard and white and high."[19]

A staff writer for the *New Yorker* magazine effused in 1981: "There are moments when this city can jolt us with the certainty that we'd rather be alive right here and now than anywhere else on earth at any time whatever." The words of praise stress the magic and gemlike glow of Manhattan in early evening, when the sun has just set and the moon hangs low in the sky, wedged between buildings, like "a great pink balloon mottled with lavender-gray." On Fifth Avenue, just outside the magazine's office,

> We had the feeling that the lights had gone up on a theatre set and that something most significant was on the point of happening. Gilding sparkled from the black frame of the Victorian facade of Scribner's bookstore, and slanting beams of brightness gave the store's depths a sacrosanct dignity to rival St. Patrick's. Every inch of glass or metal in view beamed back its rosy burnishing. The poles of traffic lights glowed like treasure, and ordinary shoe-store windows looked like jeweler's showcases.[20]

Process and Movement

America "is not an artifact. . . . America is process," says one observer of the modern American scene, which explains why, according to another, the country's most distinctive look is "casual chaos."[21] Few parts of the landscape, urban or rural, seem finished. Fences have not been put up, boundaries are left vague: towns and cities interdigitate, dribble off, or haphazardly merge with the countryside. The tools of construction—shovels, cranes, and pickup trucks—are left (it often seems) by the roadside, as though they might be used again at any time. True, Americans, like most people, also value the finished product: hence the ideal of the suburban home with its neatly trimmed lawn and their delight "in a tidy and unblemished storybook past, such as the village scene of Grant Wood's *The Midnight Ride of Paul Revere*";[22] and hence perhaps even their liking of wilderness, considered as nature's finished artwork. Nevertheless, despite this predilection, "casual chaos" in the landscape does not necessarily spell ugliness to Americans, because it has, for them, overtones of process and progress. "Unkempt" or "untidy" is a critic's negative judgment (based on visual evidence) of the *state* of a project, whereas the real source of pleasure for the pioneer farmer or builder may be kinesthetic—that is, moving things efficiently toward a goal. An orientation to the future makes one tolerant of the tacky, provisional character of a scene. Out of such rawness, Americans see with their mind's eye—a year or two down the road—a gleaming shopping mall or a trim park.

But there is also something deeply American and moving about the unfinished look itself—things standing alone, out of context, that express an as-yet-unfulfilled yearning. Why, Edmund Wilson asked, should "the sight of a single solitary street lamp on the Staten Island shore" make him pause? "It had merely shed a loose and whitish radiance over a few feet of the baldish road of some dark, thinly settled suburb. Above it, there had loomed an abundant and disorderly tree. But there was America, I had felt with emotion—there under that lonely suburban street lamp, there in that raw and livid light!"[23]

More than people in other nations, Americans have been on the move, drawn by hope rather than driven by dire necessity. Railroads played a key role in the great internal migrations, and perhaps for this reason travel by rail resonates with a mythic depth that air travel does not yet have.

You wake in a Pullman bedroom at three A.M. in a city the name of which you do not know and may never recover. A man stands on the platform with a child on his shoulders. They are waving goodbye to some traveler, but what is the child doing up so late and why is the man crying? On a siding beyond the platform there is a lighted dining car where a waiter sits alone at a table, adding up his accounts. Beyond this is a water tower and beyond this a well-lighted and empty street. Then you think happily that this is your country—unique, mysterious and vast. One has no such feelings in airplanes, airports and the trains of other nations.[24]

The beauty of American landscape seems not designed for the sedentary and the slow-moving—for those who hug the earth. It is not manifest in the details—or, if it is, then in details that cannot be seen on the ground. Interstates and freeways may be boring and the cloverleaves disorienting and dangerous at the ground level, but who will deny their monumentality and cool elegance from the air? The classical landscape of the ordinance survey is not visually obvious on the ground, but from the air it is, and this rectangular pattern, when combined with the giant circles of green produced by irrigation, is strikingly beautiful.[25] Small towns that dot the flat interior plains have little claim to visual distinction, however fondly we regard them as quintessential American places. But try observing them from the air. Fly over (as J. B. Jackson recommends) western Kansas.

It begins to grow dark and at first all one can see is a dark mottled brown world under an immaculate sky of deep blue steel; then one flies over some small rectangular pattern of scattered lights— a farm town—and out of it, like the tail of a comet, stretches a long sinuous line of lights of every color and intensity, a stream of concentrated multicolored brilliance, some of it winking and sparkling, and every infinitesimal point of color distinct in the clear night air.[26]

As one strolls and looks up from what is close at hand to take in a broader field of vision, the details fade and larger patterns emerge to form picturesque scenes of complex shapes, curved lines, and colors. Admired in Europe and along the older eastern seaboard of America are these artful aggregations of particulars—these views and land-scapes. In America, however, in the plains country beyond the hun-dredth meridian, one may look up from whatever one is doing and be confronted by abstraction—an expanse of pale blue separated from an expanse of mottled brown by a line as severe as a ruler's edge. Ab-stract nature is a part of the aesthetic experience of Americans who have settled or visited the West—a nature of sweeping horizontals and short verticals. This abstract nature of firm lines and bounded spaces is classical—indeed, Euclidean.

However, there exists another kind of abstract nature, romantic rather than classical, known to those who ride a light airplane, speedboat, sports car, or motorcycle. These machines move much too fast for their riders to observe the small creatures of nature. Even hills, woodlots, and houses become blurred images. A new land-scape emerges, composed "of rushing air, shifting lights, clouds, waves, a constantly moving, changing horizon. . . ." As yet only a few people—and these usually young and male—know personally what Jackson calls "The Abstract World of the Hot-Rodder."[27] And it may be that sensitivity to the wasting of fuel will discourage the entire ethos of speed, engaged in for fun and as a kinesthetic-aesthetic experience. If so, this is cause for mild regret. Hot-rodders hurling along a stretch of the freeway, through patches of California fog, know something of the sensual-abstract beauty of nature and the exhilaration of movement that have been part of the stock of human experience since horsemen hurled themselves across the steppes.

The Amusing,
Gigantic, and Bizarre

The child in all of us is captivated by the bright, the gigantic, the amusing, and the strange. We respond to such objects with wonder, and the anxiety that the strange and gigantic may at first arouse can turn into amusement when we see them as harmless, a joke, an exu-berance of spirit. Anything large or exceptional in some other way has the power to command attention; and attention is the beginning of respect. In the 1780s, Philip Freneau sought the respect of Europeans for his new country by referring to the Mississippi as "this prince of rivers in comparison of whom the Nile is but a small rivulet, and the Danube a ditch."[28] Giantism was one source of pride. Freakishness could also be counted on to draw admiration. Thomas Jefferson boasted of Virginia's Natural Bridge. Yellowstone's earliest advocates were enamored not so much of its wildness or natural beauty as of its natural curiosities—the geysers, hot springs, and waterfalls, which they acted to protect from uncontrolled commercial exploitation. Tourists themselves, in the early years, appreciated Yellowstone mostly because it seemed to them a gigantic outdoor "museum" in which the wizardries of nature were conveniently on display.[29]

An immense natural world that even in its moments of calm presaged violence was a fact of life to explorers and pioneer settlers as they pushed westward across North America in the eighteenth and nineteenth centuries. Humans were dwarfed. Early landscape paintings bore witness to their physical insignificance. And yet these works, even while they ostensibly deferred to nature's might, simultaneously undermined it by capturing it on canvas. The artist seemed to want to challenge nature with outsized creations of his own. Good artists did not need to resort to giantism. Less talented ones took to it to assert their own prowess and as a sure means of catching the public's notice. Enormous pictures of the West, showing extensive stretches or even the entire course of the Mississippi River, drew multitudes of viewers. The best known of these was a monster three miles long.[30]

A vast and fierce continent has provoked a boastful rather than submissive posture among pioneers. Besides panoramic canvases, other means to overcome space's tyrannous power to diminish all human accomplishments include tall tales, city boosterism, and huge roadside posters and sculptures. Hardly any communal artwork is so picayune that it cannot be elevated through flamboyant exaggeration into a cynosure of pride, a potential mecca for tourists. Aggressive commercialism, wedded to and engorged by civic pride, has encouraged the creation of a declamatory, leg-pulling popular art that is now widely regarded as distinctively American. Expressions of this art have been significantly altered by the appearance of motorcars and highway travel. In the premotorcar age, exaggeration was restricted to the verbal form of tall tales and to the visual art form of advertisement in posters, booklets, and magazines. As motor travel became commonplace, popular art responded by becoming sculptural and architectural. Flamboyance is now manifest not just in words and pictures but in three-dimensional features of the landscape.

Consider one example from northwest Minnesota. The winter of 1937 was bitterly cold, the economic climate not much warmer. Tourism had ground to a near standstill. The citizens of Bemidji needed to do something to regain their self-confidence. They put up a giant statue of the legendary worker-hero and lumberjack Paul Bunyan; beside him was his companion, Blue Ox or Babe. As motorists headed west toward town on the primary route, they were directed by these glossy red and blue figures to abandon the highway for Bemidji's first annual Paul Bunyan Winter Carnival. The figures were, in Karal Ann Marling's words, "crudely framed, garishly colored behemoths [that] demanded attention by the sheer force of their intrusion upon the flat, white wintertime landscape of Minnesota. And the 15-foot Paul, at least in his early years, attracted notice with awkward, lurching gestures produced by hidden wires, while Babe . . . cantered about the side precariously mounted on the chassis of a Model A."[31]

The Strip

American space dwarfed pioneer settlements, humbled oxcarts and horse-drawn caravans, and made walking on two feet seem impractical or foolhardily heroic. The motorcar has changed all that. Towns, once at the mercy of nature and distance, now fear being bypassed—considered unworthy of a stopover—by speeding motorists. To catch their eyes, local businessmen resort to excesses of size and color, and to creating a carnivalesque atmosphere: hence the monstrous signs that lord over the establishments they are meant to advertise, the buildings that are themselves signs (structures that look like a hotdog or an orange), and car dealerships done up with bright banners as though a jousting match on horseback were about to take place.

The strip has often been called a visual blight. Its inhuman scale, the empty spaces and parking lots between buildings that stand in ragged outline on two sides of the highway, the makeshift facades, the lurid neon lights, the unsubtle commands (EAT! GAS!), and the huge sculptural trademarks in the shape of cute humans and animals have been repeatedly criticized. Connoisseurs of art and of picturesque nature assume that aesthetic appreciation is possible only when one moves slowly, with frequent pauses, as one does in a picture gallery. Urbanologists, too, have usually taken a pedestrian's view of the streets, which can be savored only when people stroll. The American strip is entirely misread if one sees it through these traditional and high-cultural eyes. Technologically, the strip is the product of a pace of motion quite without historical precedent. It emerged in the relatively affluent post–World War II period and catered especially to a young clientele, high schoolers and recent graduates, who had found a new way to spend their leisure, cruising up and down the road, that differed radically from the leisure style of their parents and grandparents. Cruising the strip was new because it did not imitate the mores of the upper and upper-middle classes.[32] Whereas grandparents might still aspire to an afternoon by the park kiosk, listening to the band, and the parents, more glamorously, an evening in a softly lit restaurant just off Main Street, followed by waltzing to live music, the young in the postwar decades hopped into their souped-up cars and rolled down the roof and windows to enjoy the glaring lights, the shouts and honks of camaraderie directed at fellow cruisers, the pauses for root beer, jumbo burgers, and other amenities.

The strip obviously does not cater only to a local clientele. To survive, it must count on money-spending tourists and out-of-towners. The strip extends a loud welcome to strangers, who appreciate the opportunity to stop and gas up, go to the rest room, stretch their legs, and refresh themselves with iced tea or hot coffee after the long hours of driving through seemingly empty country. Much of the strip is undeniably an eyesore. The jungle of competing posters and sculptures defeats their purpose: if everyone shouts, no one is heard. And yet a passerby has to be a pinched puritan not to respond occasionally to the sheer verve of a commercialism that, at its witty best, can wink at itself.

A characteristic of other-directed architecture, J. B. Jackson says, is its dependence on signs, most of which, unfortunately, try to attract notice by mere size and blatancy. Commercial signs nevertheless have the potential to become a popular art form that appeals to more than one layer of our imagination. Well-conceived fantasy, not just size, can make tourists pause. Reassuring familiarity, such as the trademarks of various service stations or McDonald's golden arches, can make clients feel comfortable about stopping. A unique technology available to modern designers is lighting. "Neon lights, floodlights, spotlights, moving and changing lights of every strength and color—these constitute one of the most original and potentially creative elements in the other-directed style."[33] In the 1950s and 1960s civic groups and professional architects routinely attacked the flashy vulgarity of neon lights. It rarely occurred to them that modern techniques of lighting could offer artists a truly new and liberating medium. Even without the artists' contribution, the brightly illuminated strip, set against a black sky, is a fairyland to youngsters, who, unlike their harassed parents strapped in the front seat of the car, enjoy the childish directness and exuberance of the signs, unhindered either by their crudity or by subtextual messages of capitalist exploitation.

The strip is the fruit of aggressive private enterprise. Corporations and individuals own the businesses, which, because they cater to consumers, are all other-directed. This is true even of the interior spaces. Beyond the parking lots are the brightly lit or dark-cool sequestered rooms and courts that exist for no other purpose than to be constantly invaded by transient customers. Architecturally, private enterprise has produced an undisciplined, unkempt—or, from another viewpoint, uninhibited, romantic—environment. However, the strip also has a prominent public face.[34] It is the highway. In sharp contrast to the individualism run wild on either side, the highway itself is formal and classical. Its signs are austere and standard, as are also the tall and elegant aluminum light poles that stand like sentinels to ensure safety and impart a certain dignity to the unending procession of motorists.

From the Songline of Aboriginal Australia to the commercial strip in America, with intermediate stops in the European medieval cosmos and in the Chinese landscape, is an intellectual trip that reminds us of how vastly different cultural worlds can be. Time is clearly a factor in the difference. While all cultures change, some undergo more radical transformation than others. The change in cultural ways and environment in Aboriginal Australia in the millennium before European settlement (say, from 800 to 1800) was not as great as that in China during the same period. In Europe the change was greater still; and, of course, America's commercial strip owes its existence wholly to a modern capitalist economy, neon lights, and motorcars. Yet an attentive individual can appreciate and find areas of reassuring familiarity in all four cultures. Why? The fact that all humans have essentially the same biological senses and mind is no doubt one reason, but more important is how these senses work together and how the mind, in rather similar ways, builds on such biological foundations to create symbolic spaces and complex cultural worlds that have points in common. No one can say precisely how cultural worlds arise from the biological base. But it is possible to attempt a sketch of the principal building blocks, and this will be my project in the two chapters of Part IV.

IV

Synthesis of Space

and the State

Synesthesia, Metaphor,
and Symbolic Space

*T*he differences between sensation and aesthetic experience—between living unreflectively in a milieu of multiple sensations and exercising a particular sensory mode in a focused way, between the easygoing perceptions of routine life and searching appreciation—reflect the degree to which the mind comes into active play. Aesthetic experience occupies the large middle span of the continuum: too much emotion, too great an embeddedness in routine, too intense a focusing on just one sensory mode, too great an exercising of the abstract intellect will all weaken or even destroy it.

This chapter and the next explore two kinds of *multisensory* reality. Both are complex. The one is experienced by the body; the other is constructed by the mind. The one is a fact of nature or an unplanned property of the built environment; the other is a more or less deliberative creation. The world is multisensorially stimulating not only because that is how it is, but also because that is how we humans have planned it. Accordingly, we shall look first at unreflective experience and biological facts and move to increasingly complex human creations—from multimodal experience in synesthesia to the mixed modes of experience in the creative use of metaphor and simile; from the construction of symbolic spaces, which rests on a sophisticated knowledge of the powers of metaphoric language and images, to the construction (in Chapter 9) of the enormously complex aesthetic-moral state.

Multisensory Experience

Our proximate environment is experienced multimodally. What is close can be touched and perhaps tasted, smelled, heard, and seen; it has the dense texture of a reality confirmed by multiple perceptions sustained over time. With distance, however, one by one the senses lose their effectiveness: first touch and taste, then smell, then hearing, until finally only sight still informs. The perceptual field, with progressive distance from the observer, "thins": long after the nose ceases to detect odors, the ear can still discern the muffled sound of traffic; farther away, sound fades and there remains only a silent scene. The farther out the world lies, the simpler it seems and the more readily one perceives it in an evaluative-aesthetic, as distinct from "swooning," mode. Psychological distance is paralleled, in this case, by geographic distance. No wonder the visual is widely considered *the* aesthetic experience. Psychological distancing is difficult in the proximate environment because there one tends to be overcome by the senses' power to stir the emotions and also because, when several senses are simultaneously operative, it takes a special effort to attend to one effect rather than another, or to their joint complex subtleties.

Nevertheless, we are well able to appreciate certain objects in our near environment aesthetically, provided the stimuli are not overpowering. Fire in the hearth is an outstanding example. Fire is color and movement, crackling sound, fragrance, and warmth. Fire seems alive, and its ability to massage and stimulate most of the senses makes us feel alive. Yet we can also withdraw from it sufficiently to contemplate it as an object. From the depth of an armchair, with the lights dimmed and the curtains drawn, we can luxuriously toast ourselves before a burning log, inhale its scent, listen to its comforting crackle, and gaze at the dancing shapes of the flames.

Unlike hearth fire, the countryside is not just an element in the environment. It *is* the environment, and it too offers multisensory stimulation. If the countryside is unfamiliar to us, we are likely to attend primarily to its visual aspects; if it is familiar or ordinary, nonvisual qualities may come to the fore. John Cowper Powys puts it thus:

> Suppose you are spending your next free summer afternoon taking a leisurely walk through an ingratiating but quite undramatic landscape. What you will be at once aware of, and aware of far more intensely than any particular colours or forms or sounds, will be what I describe as the smell of the day. In the next place, and before you concentrate upon any particular objects, you will be vividly conscious of the touch of the day. By this I mean the warmth or the chill of the atmosphere, the exhalations of cold or heat from the actual soil under your feet, and, above all, the feel of the wind on your bare skin. But now in the third place we come to the crucial point of all: namely the taste of the day.

What Powys means by "taste" is the combined appeal of the senses, or, in his down-to-earth words, "chewing the cud of sensuous satisfaction."[1]

Synesthesia and Synesthetic Tendency

Synesthesia is a curious physiopsychological response that is only partially dependent on external stimuli. Synesthesia occurs when the stimulation of one sensory mode (such as taste or smell) activates another (such as hearing or vision). A common form of synesthesia is "colored hearing," a condition wherein the hearing of a sound induces the visualization of a color.[2] Vowels in human speech evoke colored images with remarkable consistency. Still more common is the association of the pitch of a sound with the brightness of an image. For example, low-pitched sounds such as deep voices, drums, and thunder produce dark images, whereas squeaks, violins and soprano voices produce white or bright images. Another common association is between the pitch of a sound and the size and shape of an image. Thus high-pitched sounds are small, angular, and sharp-edged, whereas low-pitched sounds are dark, round, and massive.[3]

Synesthesia can be highly individualistic and specific. One person informed Francis Galton that to him the letter A was always brown. To the poet Arthur Rimbaud, on the other hand, A was black.[4] Vladimir Nabokov, as a child, was endowed (or afflicted) with a high degree of sound-color synesthesia. The long a of English had for him "the tint of weathered wood, but a French a evokes polished ebony."[5] To a Russian journalist, every sound produced an experience of light and color, and at times also of taste and touch. Listening to a man speaking, he said, "What a crumbly, yellow voice you have." Synesthesia enhanced the journalist's memory. Every object he encountered was etched deep in his mind at least in part because, although only one sense was stimulated, the others were also activated to reinforce the impression. As the journalist left the institute where his memory was tested, a scientist unthinkingly remarked, "You won't forget the way back to the institute [on your next visit]?" "Come now," said the journalist, "how could I possibly forget? After all, here's this fence. It has such a salty taste and feels so rough; furthermore, it has such a sharp, piercing sound. . . ."[6]

True synesthesia is rare; but synesthetic tendency—the ability of a stimulated sense to evoke a mood associated with another mode of perception—is commonplace. Nabokov's claim that the French a evoked "polished ebony" may strike us as poetic fantasy or as evidence of a curious psychological disorder. On the other hand, most of us see nothing strange in associating low-pitched sounds, like those of a drum or thunder, with dark images, and high or squeaky sounds with white or bright images. Nor do we think it strange that high-pitched sounds are somehow small and angular, whereas low-pitched sounds are massive and round. True synesthetes will actually *see* a dark image—a dark shape—when they hear the roll of thunder. The rest of us will merely feel that in some inexplicable way the sound affects us with a "dark and massive" feeling.

Charles Osgood has shown that synesthetic tendencies are widely shared across languages and cultures. In a study of Anglo-American, Navaho, and Japanese subjects, he found that all of them perceived the idea "fast" as thin, bright, and diffuse; "heavy" as down, dark, and near; "quiet" as horizontal; and "noisy" as crooked. The world's "feeling tone" seems to be similar everywhere, regardless of cultural and linguistic differences.[7]

We do not yet understand why people have transcultural synesthetic tendencies. Part of the answer may lie in basic shared physical experience. According to Osgood, "[It] is simply a characteristic of the physical world that as a noise-producing object approaches or is approached, increases in visual angle are correlated with increases in loudness."[8] And part of the answer may lie in neurophysiology. Whatever the cause or causes, we experience the world in terms of feeling-tones, without which human reality would be greatly impoverished. The texture and vibrancy of the world are thus a consequence of synesthesia. Yet the synesthetic experience is not itself aesthetic. We do not pause to wonder at the evocation of "dark roundedness" with the boom of the drum; the blending of the senses is so habitual that we take it completely for granted. Nevertheless, that mental capacity called synesthesia and synesthetic tendency is the primitive ground from which emerge metaphoric perception, the transforming and evocative powers of language, and (generally) human understanding and appreciation.

Metaphor

A synesthete may say something that sounds poetic, such as "What a crumbly, yellow voice you have," and yet be a very literal-minded individual, with little understanding of metaphor and poetry. It is often assumed that poetry calls for a highly graphic kind of imagination. In fact, the metaphors in poetic language evoke not so much images as moods or atmospheres, and ideas. We say of a lovely young woman that "she is a rose." An extreme synesthete like the Russian journalist may well find the expression nonsensical. Put a rose next to a woman, and where is the resemblance? A rose looks much more like a cabbage than like a young human female, and yet adults in Western culture have no trouble understanding that figure of speech.

Young children tend to be more synesthetic than adults. A biological advantage of this propensity is that, by making objects vivid and easier to remember, children find it easier to gain confidence in the substantiality and reality of the world. As they grow older, they depend less on the services of synesthesia and more on the flexible resources of language. Children's metaphors tend to be more perceptual than ideational. Young children, says Howard Gardner, may describe nuns as "penguins," but they do not find it easy to understand psychological-ideational metaphors such as "a heart of stone," or the comparison of love to a summer's day.[9] In time, as young adults, they will understand, and some may invent splendid figures of their own.

Language is chockful of quiescent metaphors, and the people using them are unaware of the way they originally extended and enriched meaning, or of the subliminal effect they continue to have on perception and action. People who believe themselves to be thoroughly prosaic in their use of language are in this regard like synesthetes who take the multisensory quality of their experiences for granted. A world dulled by habitude can be redeemed by fresh metaphors—new ways of seeing, feeling, and understanding. Encountering them, or creating them ourselves, we recognize, paradoxically, both their newness and their relevance to enduring aspects of our being. A new metaphor, though clearly the product of someone's mind, seems (if it is good) not an invention at all, but rather a gift—an intrusion of reality.

Certain habits of seeing are widely shared. "Earth mother" and "sky father" are instances, as is also the understanding of the earth in anatomical terms: rocks are bones, soil is flesh, and the plant cover the hairpiece of a cosmic being. The English-language names of many landforms incorporate such anatomical references: examples include headland, foothill, volcanic neck, the shoulder of a valley, and the mouth of a river. Animal epithets for human beings are probably universal. We seek to capture the essence of people's personalities or behavior with reductionist epithets such as "catty," "bullish," or "piglike."[10] It hurts to be called a pig, and it is no good my trying to console myself by saying, "But that is just a cliché—a tired metaphor." Certain metaphors, for good or ill, appeal to something deep within us and have enduring power. To call someone a certain kind of animal is always likely to evoke an emotional-aesthetic response, although the meaning of a particular animal, and thus the intended response, may differ from culture to culture.

Colors, too, have widespread metaphorical uses. Some appear to have universal meaning based on synesthesia or common experience: thus red signifies warmth and energy, light blue coolness.[11] Some convey different meanings depending on the culture in which they are used: thus yellow is an imperial color of high prestige in China, but it seems loaded with negative connotations in the West (as in "You are yellow" and "yellow press"). Still others carry different meanings depending on context: "You are green" may be taken to mean "You are naive," or "You are sick," or "You are jealous."

Metaphors of location appear to transcend culture: "I am central" can only mean "I am important—someone to be looked up to"; "I am peripheral" is a humbling admission in any language.

Symbols and Symbolic Space

Metaphor reaches backward into synesthetic tendency and forward into symbol: in one direction it becomes an automatic response; in the other, a product of culture and the active imagination. Symbols are evidence of the human capacity to let one thing stand for another. They may be as abstract and impersonal as "Let x be . . . ," or as tangible and emotionally resonant as the Christian cross or a nation's flag. Understanding of these abstract or tangible signs depends on a rich background of cultural elaboration. In contrast to the metaphor, which may couple unlike entities in a flash of inspiration, illuminating both without the necessity of further thought, the symbol is the consequence of an extended excursion in analogic reasoning. Plato accepted the microcosmic metaphor of his time—the idea that the human body is a likeness of the cosmos. He did not, however, rest with the metaphor; he proceeded to build a towering schema of correspondences harmonizing the components of the universe from the very small to the very large and, in the process, transformed the human body into a multilayered symbol.[12]

Symbolic space offers good examples of how the human imagination works. Space becomes symbolic when it intimately conjoins human and social facts with those of nature.[13] Symbolic space is a mental artifact, necessary to the ordering of life, and so in this sense it is a practical venture; and yet it is also infused throughout with the aesthetic values of balance, rhythm, and affect. Symbolic spaces have different foundations and exist at different scales. The space based on a grid of center-periphery and the cardinal points, its scale ranging from a small object (such as the sacred pipe) to a large country, seems to be deeply congenial to human practice, mind, and imagination, for it occurs worldwide—in the New World, Africa north of the Sahara, Europe, the Middle East, and Asia. Wherever it occurs, the spatial frame of center and cardinal points is loaded with other symbols, which may include combinations of color, animal, seasonal or meteorological phenomena, and human social categories and activities.[14] The degree of elaboration varies from society to society: larger and more complex societies tend to have the more highly articulated and embroidered spaces. The degree to which individuals can stand sufficiently outside their symbolic space to appreciate its scope, meaning, and beauty also varies widely. Within each culture, only a small minority of people, who function at some level as teachers, feel at ease verbalizing the group's values: they are, depending on the culture, the shaman, the medicine man, the courtier, or the scholar. Yet all members of a group who participate in its ceremonies may be presumed to have a *feeling* for what is going on, such as praying in the direction of the rising sun; and this feeling is tinged by a sense of appropriateness, of something that is done satisfyingly right, of blessedness and beauty.

Symbolic space, though created for a variety of purposes, is also rightly seen as the embodiment of strong aesthetic impulses—an artwork. Let us look first at the symbolic spaces of two traditional societies at opposite sides of the earth—the Oglala Sioux and the Chinese; then we can consider whether the type of symbolic space that is at the heart of many premodern societies also exists in the mythology of modern secular states such as Australia and the United States.

THE SIOUX NATION AND CHINA

Black Elk, a holy man of the Oglala Sioux, said to John Neihardt in 1931:

> I will first make an offering and send a voice to the Spirit of the World, that it may help me to be true. See, I fill this sacred pipe with the bark of the red willow; but before we smoke it, you must see how it is made and what it means. These four ribbons hanging here on the stem are the four quarters of the universe. The black one is for the west where the thunder beings live to send us rain; the white one for the north, whence comes the great white cleansing wind; the red one for the east, whence springs the light and where the morning star lives to give men wisdom; the yellow for the south, whence come the summer and the power to grow.

All the essential elements of the Sioux symbolic space are captured in these remarks. Each of the cardinal points is associated with a color and a powerful being or force of nature. Under the spell of the holy man's eloquence, even an outsider can feel the rightness and drama of black clouds bringing fertilizing rain from the west, the clean, harsh beauty of snow-laden wind from the north, the red sun that heralds dawn in the east, and the yellow heat of noon and south that promotes growth. Oglala space is a sacred hoop. At its center is a tree in blossom—the axis of the world. There, the spirits bring one to see "the goodness and the beauty and the strangeness of the greening earth, the only mother."[15]

The Oglala sacred hoop is also a picture of time—the round processes of nature. Again, Black Elk draws the separate elements of nature and people together in vividly figurative language:

> The sky is round. . . . The wind, in its greatest power, whirls. Birds make their nests in circles, for theirs is the same religion as ours. The sun comes forth and goes down again in a circle. The moon does the same, and both are round. Even the seasons form a great circle in their changing, and always come back again to where they were. The life of a man is a circle from childhood to childhood, and so it is in everything where power moves. Our tepees were round like the nests of birds, and these were always set in a circle, the nation's hoop, a nest of many nests, where the Great Spirit meant for us to hatch our children.[16]

Symbolic space, in sharp contrast to geographic space, transcends scale: the small is in its own way as complete an evocation of mythic reality as the large. To the Oglala Sioux, a pipe with four colored ribbons, the bird's nest, a single tepee, a number of tepees arranged in a hoop, and the vast space circumscribed by the course of the sun may all carry the same basic symbols and evoke the same richly articulated and emotionally satisfying world.

The Chinese symbolic space, though far more complex than that of the Oglala Sioux, shares with it certain fundamental characteristics. These include a spatial frame defined by the cardinal points and "the round processes of nature," which in the Chinese view function under the guidance of the grand cosmic principles of *yin* and *yang*. To the Chinese version of the spatial frame and cosmic clock are attached numerous other components, including, prominently, the five elements, the five colors, the five animals, and the five offices. The outcome, elaborated over the centuries, is a vast spatiotemporal edifice, imbued with moral-aesthetic overtones, that has had extraordinary power to give reality a sense of coherent form and value for millions of people in a country the size of a continent. And it has been able to do so until modern times. Indeed, elements of the edifice persist to this day. We can look at it from different angles and argue that it has served a number of distinct purposes—sociopolitical, religious-ceremonial, and moral. But whichever aspect we choose to emphasize, the fact remains that it is also an outstanding work of the imagination, a long-enduring artwork of pervasive influence.

In symbolic space, cardinal points are neither points nor, strictly speaking, directions. Rather, they are directional regions. East is the region of sunrise and spring, the color green, the element wood, the animal dragon, and civil service. South is noon and summer, red, fire, phoenix, and imperial augustness. West is sunset and autumn, white, metal, tiger, and military service. North is night and winter, black, water, reptilian animals, and commerce. Center is the region of the yellow earth and of man. This is the barest characterization. Hundreds of other elements have been built onto the frame over time. Many are of local origin, have only local interest, and may be transient. Prompted by the needs of particular times and places, these local efflorescences are readily accommodated within a general structure that is authoritative throughout the Chinese world.[17]

Symbolic space is a different sort of mental construct from geographic space. Yet, because both are grounded in experience, they have areas of overlap. Symbolic space is geography elevated and transfigured. It is the common earth inscribed with the sublime simplicities of heaven and the rounds of socioeconomic life illuminated and enriched by poetry. Why those colors in those particular directions? According to some scholars, nature provides part of the answer. Yellow is the color of loess in the upper Yellow River (Huang Ho) basin—the cradle of Chinese civilization. White is the snow on the high Tibetan plateau, green the fertile plain and the blue-green sea to the east, and red the oxidized soils of subtropical China.[18] The colors, however, also reverberate with meanings derived from heaven and human affect. Green is indeed the color of vegetation, but, more generally, it stands for growth and life; as such, green merges with sunrise, spring, and civil service, the human institution that promotes and regulates life. The meaning of the region East is enriched by mutually supportive figures drawn from the three realms of heaven (sunrise, spring), earth (wood), and man (civil service). The ways in which these figures merge with and reinforce one another are exemplified by the blue-green dragon, a mythical being that belongs to both heaven and earth, and moves along an axis that links the two.

The other regions likewise pulsate with meaning, and for similar reasons. Thus South is high noon and imperial, fire, and red; and red is more than just a reference to the color of soil; it is a "high" color, an auspicious color, the color of blood and life. West is sunset and autumn, cool metal rather than the fire of summer. Autumn is the season of declining warmth and life, and yet it is also the time of harvest and (potentially) of abundance. The word for autumn, *ch'iu*, is made up of "grain" and "heat," the heat that produces an abundant harvest of grain. Thus it seems that the fire and life of the South have carried over to the fall. Nevertheless, the predominant sense of West is one of waning. White may be the color of snow, but even more, it is in China the color of mourning and of death. A premonition of violence and death is heightened by the West's animal emblem, tiger, and the region's linkage with military service. North is darkness, water, and reptiles. It suggests the unformed or primordial; it is the profane region of commerce, and yet the triple figure of darkness, water, and hibernating reptiles also connotes renewal, fertility, and life.

The round processes of nature appear in China as the alternation of the two cosmic principles of *yin* and *yang*. The ascending (*yang*) seasons of spring and summer are followed by the descending (*yin*) seasons of autumn and winter. The five elements, too, follow or produce each other in sequence: thus, as Tung Chung-shu (ca. 135 B.C.) put it, wood produces fire, fire produces earth (that is, ashes), earth produces metal (that is, ores), metal produces water, and water produces wood. "This is their 'father-and-son' relation. . . . As transmitters they are fathers, as receivers they are sons. There is an unvarying dependence of the sons on the fathers, and a direction from the fathers to the sons. Such is the Tao of Heaven."[19] Symbolic space, read temporally, is a cosmic clock that registers the cyclical processes of nature to which humans and human institutions must conform, ritually and in the mundane practices of agriculture. Yet humans are at the center of the cosmos, and by their acts of commission or omission they can enhance or disturb the harmony of the universe. Supremely, this power resides in the emperor, perceived as representative man, father of his people, and Son of Heaven, that is, a quasi-divine being. The emperor mediates between heaven and earth. He occupies the center, which also connotes height. The emperor, from on high, looks southward; West is to his right, East to his left, and behind him is the profane region North.

Even this brief sketch of Chinese symbolic space reveals it as a magnificent construction of thought and feeling. It plays a role not only in the rituals of state and in lesser ceremonies, but also in the practical operations of life, including agriculture and the design of cities, palaces, and ordinary houses. The Chinese occupy this symbolic space as part of their everyday existence. At the same time, people can exercise observation and critical appreciation as they follow the rituals of symbolic space and even as they live in a house and a city informed by its principles.[20] Because the metaphors and symbols are age-old, they have undoubtedly lost much of their power to surprise, except to the young who are initiated into them every generation. But this does not mean that they are incapable of arousing aesthetic pleasure.

AUSTRALIA AND THE UNITED STATES
Symbolic space based on the cardinal points, though widespread in traditional cultures, is not universal. Dwellers in the tropical rain forest do not have it, if only because they can barely see the sky and the sun. Australian Aborigines live under the glaring sun and an overarching sky, but their symbolic space is built on a narrative-historical structure of paths or Songlines rather than on a grid of cardinal points. The cosmic-symbolic mode of thinking has lost much of its potency in modern times; nations and cultures established since the eighteenth century have been much more concerned with gaining power and control through geographic knowledge than through the construction and refinement of symbolic space, with its attendant rituals, ceremonies, and art.

Consider Australia. Its eastern portion is named Queensland, New South Wales, and Victoria; the rest of the country bears the directional labels Northern Territory, South Australia, and Western Australia. These labels, however, carry no poetic resonance; they seem to indicate merely that government officials quickly ran out of inspiration. The "Western" of Western Australia does not in the least signify twilight, the fall season, romantic frontier, or Avalon. "Northern" and "South," whatever metaphorical resonance might be built into these words, are contravened by antipodal experience—the reversals of heat and cold, light and darkness on the southern continent. Nevertheless, since 1800 Australia has gradually developed its own symbolic space, superimposed on the far older Aboriginal one. This later symbolic space of European imagination is a construct of center and periphery rather than one of cardinal points, and it reflects the location, abundance, and economic value of the continent's resources: the core, arid and incapable of supporting a large population, is little valued in comparison with the periphery, where both rainfall and people are concentrated. From a mythic point of view, the relative importance of core and periphery is reversed. The core is the country's sacred space. Words for the core, such as *heartland*, *brown country*, *upcountry*, *bush*, and *outback*, with positive connotations, outnumber negative words such as *dead heart* and *desert*. Even when the emphasis is on the desolation of the center, a certain mysterious grandeur—the terror of the sacred—is implied, reinforced by epithets such as "primeval" and "timeless." But in a profound sense the heartland or bush country was also home (sacred homeland) to pioneer Australians, their writers, and mythmakers.[21] The real Australia, in the literary imagination, is not the periphery, and certainly not the large and growing cities with their increasingly cosmopolitan culture, but the interior.

The United States, too, has its heartland mystique. In the American version of the myth, authentic America is neither the East coast, with its European connections, nor the West coast, with its propensity for surreal fantasies; rather, it is the solid Middle West. Even today foreign dignitaries are expected to visit an Iowa farm if they want to claim that they have seen the real country.[22] Unlike Australia, the core area of the United States is exceptionally fertile. The heartland mystique thus does not conflict with geographic reality. But it did have to compete with the rival mystique of the West, a term that eventually embraced half the continent. The core became the Middle West, a curious hybrid, rather than simply the Middle. Ohio country was originally the West. By the late 1860s, Americans east of the Mississippi and north of the Ohio began to drop the term West for their part of the republic in favor of Middle West or Midwest. Early in its history, this vaguely defined area of great agricultural wealth and of rapidly growing cities claimed to be not only the economic but also "the moral and social epicenter of the nation."[23] Midwestern congressmen agitated briefly for moving the national capital inland to St. Louis. After decades of competition for supremacy among Cincinnati, St. Louis, and Chicago, by the end of the nineteenth century Chicago was the undisputed transportation and manufacturing pivot of the heartland, and into the twentieth century it also aspired to become the literary center of the United States and the capital of a specifically American civilization.[24]

Like Australia, the United States as a whole is divided into regions with directional labels. However, unlike those in Australia, the regions designated by cardinal points in the United States have not been promulgated by a central authority; and like those of symbolic space in premodern societies, the names and meaning of American regions have been acquired over time, as part of the growing lore of a people. In the symbolic spaces of the Oglala Sioux and the Chinese, cardinal points are tied to astronomical events and to the seasons with their control over life and death. American space is not a stage set for the enactment of cosmic drama, but, as regional novels and literature show, nature—and particularly climate, controlled by the sun—does play a large role in giving personality to regions.

From the earliest times, according to Leslie Fiedler, American writers have tended to define their country in terms of the four cardinal points. "Correspondingly, there have always been four kinds of American books: Northerns, Southerns, Easterns, and Westerns, though we have been accustomed, for reasons not immediately clear, to call only the last by its name." The Northern is "tight, grey, low-keyed. . . . Typically its scene is domestic, an isolated household set in a hostile environment. The landscape is mythicized New England, 'stern and rock-bound,' the weather deep winter." The Eastern links America with the Old World in a swirl of flirtation and romance. The "season is most appropriately spring, when the ice of New England symbolically breaks and all things seem—for a little while—possible." The Southern, in contrast, "seeks melodrama, a series of bloody events, sexual by implication at least, played out in the blood-heat of a 'long hot summer.' " The Western is confrontation with the Indians, cattle rustlers and other bad guys in an overpowering and alien landscape. In its most sanitized form, it is the cowboy myth of popular fiction and movies in which the lone hero, after saving the homesteaders, rides into the sunset.[25]

When the Oglala Sioux and the Chinese compare notes on their respective symbolic space, they can discern, despite numerous and important differences, a common feeling-tone, based on the way the cardinal points in both cultures evoke a season or a type of weather, a color, a mood, or a human quality. American symbolic space is vastly different from that of traditional societies, and yet they overlap unexpectedly. American space, too, has its mythic heartland and periphery and its regions designated by cardinal points. Each region is enormously complex in physical environment and human personality; yet in the mythic imagination it appears to have only one season, one type of landscape, one color and the mood it projects, and one simplified image of society.

The basic metaphors and symbols of the Oglala Sioux and the Chinese have been remarkably stable. However, the power of metaphors and symbols can erode over time as new ways of living, new experiences, and new values emerge. American symbolic space, derived essentially from nineteenth-century experience and literature, is already slipping rapidly into the past: North, rather than evoking deep winter, may call up a pleasing image of summer as New England becomes a resort for tourists: and a modern South, with its air-conditioned cities, is less and less suitable as the stage of strong passions, played out in the long hot summer of the countryside. As modernization proceeds, American space may still be beautiful and regionally varied, but it will be a different kind of beauty and variety, for it will lack the amplitude of emotional resonance that only a deep engagement with nature and the rhythm of the seasons can produce.

CHAPTER 9

Ritual and the Aesthetic-Moral State

Society cannot function without its systems of symbols, of which symbolic space, territorially organized, is one. Here I would like to examine the ways in which symbolic space is integrated into a larger and more complex whole—the state. The state is an aesthetic-moral construction on a grand scale. Politics lies at its core, but politics viewed as the serious effort to articulate excellence for society is itself a moral-aesthetic aspiration, and its achievements are properly deemed artworks that make claims to being both beautiful and good. Rituals and ceremonies sustain the state, even a modern one. Have they always had a strong aesthetic thrust? How have they changed over time—how, for instance, do ancient rituals differ from modern civic ceremonies?

A primary purpose of ritual is to place the individual, more often the group, in a larger setting; and if this larger setting includes not only society and nature but also supernatural elements, then the ritual is considered religious. Religious rituals are comprehensive, embracing transcendent cosmic elements. Civic ceremonies, in contrast, operate essentially on a horizontal plane; supernatural and numinous elements, if invoked at all, are invoked perfunctorily.

Rituals introduce a firmness of line, shape, and procedure normally absent in daily life; they also heighten sensation, imparting color and vividness to ordinary reality. Thus considered, rituals are clearly aesthetic. However, not all are aesthetic to the same degree. They are marginally so if, at one extreme, they assault the senses and emotions or, at the other extreme, if they are too predictable. All rituals risk becoming routine and boring. The inclusion of violence, danger, and the sexual act forestalls that possibility but also diminishes the psychological distancing necessary to aesthetic experience.

Violence,
Mockery, and Degradation

Violence, involving bloodshed, was a fairly common element of ritual acts in the distant past and among certain primitive peoples even in the early part of the twentieth century. The sacred demanded blood; some scholars believe that violence and the sacred were inseparable. Among civilized societies in modern times, violence and sex have been curtailed in public rites and ceremonies. They still exist, of course, in the arts, but they tend to be confined to the cooler media of text, pictures, and sculptures.

The pre-Columbian Aztecs in Mexico furnish a gruesome example of the use of violence in ritual. Their beliefs and actions show that power over nature is compatible with a feeling of cosmic insecurity. The Aztecs had the technical and organizational skills to build architectural monuments and a complex civilization, yet they felt so little reassured by their accomplishments that they sought to guarantee the constancy of nature with bloody human sacrifice. The sun, although it rose every morning, could not be counted on to do so without propitiatory offerings of human blood. Nourishing the earth and the sun, "our mother and our father," with human lives was the first duty of man. To shirk that duty was to betray not only the gods but all humankind.[1]

Less well known to us than the Aztecs' sacrificial rituals are the cruel practices among the ancient Chinese, particularly during the Shang dynasty (ca. 1500–1030 B.C.), who also sought to maintain the fertility of the earth by means of human sacrifice. "One gains the impression," Wolfram Eberhard writes, "that many wars were conducted not as wars of conquest but only for the purpose of capturing prisoners."[2] The prisoners were killed and offered to the gods. In the feudal courts of the Early Chou dynasty (ca. 1030–722 B.C.), an exorcist danced to inaugurate the New Year, and the ceremony ended with the quartering of human victims at the four gates of the principal city. Although ritual murder was prohibited by the sixth century B.C., it continued to be practiced in the remoter regions whenever natural disaster threatened. During a severe drought, social outcasts (witches) might be forced to dance in the fields until exhausted and then be burned to death. Ritual violence declined as manners in society as a whole softened. By the second century B.C. only effigies, not real human beings, were used in the officially sanctioned sacrifices.[3]

Historical and ethnographic literature is rich in gory descriptions of fertility rites. Even the bloodiest and most openly sexual among them included certain aesthetic elements such as the shape of the altar, the color of the costumes, the processions, chants, and dances. But in their totality these rites were not intended to be aesthetic, nor should we view them as such now. Culture and the aesthetic overlap, but they are not identical. Violence and sexual exuberance in fertility cults were actions undertaken to ward off disaster, rather than dramaturgic representations. In these actions, emotional involvement and identification were maximal, distancing and observation minimal.

Besides bloody violence, the historical and ethnographic record on sacred rites also reveals mocking and obscene behavior that seems bizarre and out of place to people raised in gentler cultures and times. To Chinese scholar-officials and cultivated Europeans of an Apollonian or puritanical bent, ritual came to mean propriety, the indirection of symbolic acts, and seriousness. All the cruder signs of physicality (visceral passions and biological processes) were systematically repressed. No such desire to rise above the urgencies of the body was evident in the early stages of these civilizations; nor is such desire a powerful factor in the festive conduct of people in the lower ranks of society. Among certain nonliterate groups, too, anarchic behavior—mockery of all social conventions, including the sexual—may find an integral place in ritual. Consider the Pueblo Indians of the American Southwest. In the annual rites that celebrate and propitiate nature, a high sense of propriety prevails, reflecting a belief in an ordered moral universe that responds to prayers if the petitioners are pure. In equinoctial ceremonies, the ideal of moderation and balance is conspicuously on display. To the Hopi, the presence in the dance of the elaborately dressed ancestral deities (or *kachinas*) is a reminder that if they participate scrupulously in the rites of supplication life will continue in abundance and harmony. To the Zuñi, the presence of the Koko (rain and sky gods) is especially reassuring, for the Koko, who wear masks of intricate design representing the building blocks of the world, stand for an achieved order. Yet chaos and obscenity are also admitted to Zuñi ritual. Functioning against the world-sustaining Koko are the Newekwe dancers, who enact the precultural life of uninhibited, antisocial behavior. They may chew pieces of dirty rag and eat the human and canine excrements that have accumulated in the plaza. They are deliberately offensive—sexually and in other social spheres; they can be destructive, axing ovens and other property they encounter.[4]

The Newekwe are clowns who mock. Other Pueblo also have clowns. Indeed, the laughing, mocking clown who erupts into and transforms a solemn ceremony is a popular figure in many cultures. In medieval Europe, for example, he was to be found at carnivals and feasts, ever ready to jeer at pretensions to stability, harmony, and the elevated moral ground. His often grotesque behavior sought to degrade "all that is high, spiritual, ideal, abstract." He strove to bring high culture to the material level of earth and body—to dirt and filth, to the stomach and the reproductive organs, to acts of defecation and copulation, but also to those of conception, pregnancy, and birth.[5]

What role do mockery and other forms of degradation serve in ceremonial events? Several interpretations are possible. One is that they are a warning against hubris. Focusing on the lower stratum of the body powerfully reminds humans of their precultural and preaesthetic animal status: it proclaims that nudity and rags, rather than bright feathers and carefully decorated masks, are the human condition. A second interpretation is that the derisive gestures challenge the idea of static, permanent achievement. Art—and perhaps all serious human endeavor—strives for the stability of perfection: the perfect poem is one in which not a word can be altered without doing it harm, and, in a more practical realm, a perfect way to chop wood deserves to be handed down unchanged from generation to generation. An attack on the claim of permanent achievement may be a wish to guard against hubris, but it may also be driven by the sense that, whereas perfection implies termination, a kind of deadness, mocking laughter breaks up the rigidity, and so makes room for the new. By this third interpretation, mockery and derision signify the powers of regeneration.

The Chinese Aesthetic-Moral State: T'ang Empire

A righteous scholar-official in Confucian China would have countenanced none of the interpretations offered above. Nakedness, references to the bodily processes, crudity, confusion, violence, and destruction signified barbarism or a society in decline. True, humans had bodies and were sexual beings, but to be human was to rise above one's animal state, to refine and educate the senses and the passions so that one could live with others in a world of harmony, beauty, and virtue—in a world regulated by rites and social etiquette. An enlightened scholar or even an enlightened ruler (such as the T'ang emperor Tai-tsung) was suspicious not only of the biological furors but also of the ultrahuman realm of spirits, gods, omens, and demons. Early Chinese literature (such as the proto-Confucian *Book of Documents*) is remarkable in its lack of mythological dramas—those interpenetrations in the affairs of gods and men—that are such a striking feature of the Homeric epics. By the fifth century B.C., Chinese moralists had already removed much of the violent, the ecstatic, and the mystical from the rites of government in favor of "the image of an all-embracing, sociopolitical and cultural order in which men relate to each other in terms of a structured system of roles—familial and political."[6]

Modern scholars tend to see this order as a device to enforce the privileges of hierarchy. Rituals are the most sophisticated means of exercising power.[7] But this view, if unthinkingly applied, distorts social reality and human intention. Power has different meanings and manifestations, not all of which are bad. Power is physical energy, unpredictable as sudden shifts of the wind and predictable as the daily march of the sun; it is both the rightful exercise of authority, on which the orderly transactions of social life depend, and its abuse; and power is creativity, the energy that calls into being the aesthetic forms of culture. Throughout their millennia of imperial history, the Chinese strove for form—for cosmic harmony and the proper enactment of rites, which, in their view, was synonymous with civilization.

Chinese symbolic space, as we saw in the preceding chapter, is an enduring cultural-aesthetic achievement. But it is far more than a mere verbal-literary account; it is also architecture, city planning, and ritual. And its aesthetic-moral aim is central to the endeavor of conceiving and building a cosmos suited to the dignity of its human inhabitants.

Preserved in the *Book of Rites* is a brief description of how the capital city should be laid out. "The capital city is a rectangle of nine square *li*. Each side of the wall has three gates. . . . The Altar of Ancestors is to the left [east], and that of Earth, right [west]. Court life is conducted in front, and marketing is done in the rear." The city plan is a model of the Chinese symbolic (cosmic) space, with man at the center. The emperor sits on the throne in the audience hall at the center of his capital. The center is not only a position on horizontal space but also connotes height. The emperor faces south and looks *down* the principal north–south avenue (the axis of the cosmos) to the human world. In an imperial audience, civil officials enter the courtyard from the east, military officials from the west. The emperor has his back to the north, which is profane space and a region of darkness. And it is there that the market should be located.[8]

Remarkably, when China was reunited in A.D. 589 after several centuries of disunion, the founder of the new (Sui) dynasty decided to construct his capital in accordance with the ideal plan described in the literary canons. The city rose from the ground with a monumental rectangular wall of large circumference, oriented to the cardinal points, with three gates representing the three months of the season on each of the four walls. The main south gate of the palace–administrative city was named Vermilion Bird—a *yang* symbol of solar energy. The principal gate along the east wall was called Brightness of Spring, the direction of sunrise, the origin of the new warmth of spring after the darkness and cold of winter.[9] Although several departures from the ideal plan proved necessary, the final result was nevertheless a recognizable cosmic diagram. The Sui dynasty lasted barely three decades, to be followed by the glorious T'ang dynasty, which endured for three centuries (618–906); and it was as Ch'ang-an, capital of T'ang dynasty, that the city became the great cosmopolitan center and architectural wonder known throughout Asia, inspiring copies (Nara and Kyoto) in eighth-century Japan.[10]

The *Book of Rites* also preserves a ritual centered on a building called the Ming-t'ang, or Hall of Light.

> In the first month of spring, the Son of Heaven occupies the apartment on the left of the Ch'ing-yang hall; rides in the carriage with the phoenix bells drawn by the azure-dragon horses and carrying the green flag; wears the green robes. He eats wheat and mutton, gives orders for sacrifice to the hills and forests, the

streams and lakes. In the first month of summer, the Son of Heaven occupies the apartment on the left of the Ming-t'ang; rides in the vermilion carriage, drawn by the red horses with black tails and bearing the red flag. He is dressed in red robes, and wears the carnation jade. He eats beans and fowls, and entertains his ministers and princes with much observance of ceremony and music. . . .

And so on to other points of the compass in other seasons. The ruler's circumambulation of the rooms of the Ming-t'ang in the course of the year showed his submission to the order of space and time, but it was also a demonstration of his power—of his role in cosmic maintenance. The emperor's gestures, clothing, and food, by conforming ritually with the year's twelve months and five seasons, ensured harmony among the myriad components of the universe. At the same time, the cosmic energy thus released was absorbed by the ruler and augmented his potency.[11]

Such, in broad outline, was the ancient paradigm. In the T'ang dynasty, different versions emerged of just what the ritual meant and how it was to be conducted. Strong advocacy by one faction or another repeatedly frustrated the second and third emperors' efforts to construct the Ming-t'ang. An imperial librarian urged T'ai-tsung, the second emperor, not to waste time fretting over the size of rooms and the number of windows or doors. "If you allow the Confucians to hold different views without quickly arriving at a solution, it will only delay the performance of your rites." Early accounts of the Ming-t'ang suggest that it was a structure of several rooms, topped by a humble thatched roof. Modern scholars believe that the building's bareness accommodated the idea of passage from death to rebirth, which occurred symbolically with each movement of the emperor toward another compass point. A rather austere beauty and seriousness probably imbued the original physical setting and rites. In time, however, spectacle and magnificence moved to the fore. The fact that one could choose among conflicting versions of the rites made them seem more a human venture than pious submission to the governing powers of the universe. When the Ming-t'ang was finally constructed under the aegis of the Empress Wu in the late seventh century, it rose as a large building 300 feet high, decorated with gold, pearl, and jade and an ornate roof.[12] By then, the Ming-t'ang's cosmos-sustaining role was subordinated to the secular end of legitimating and sustaining political power.

In most traditional and folk societies, ritual is directed at the gods and owes its numinous aura to their presence. In China, by contrast, ritual tended to be viewed in human rather than divine terms. Ritual upheld the cosmos. At the same time it was seen as satisfying a people's deep emotional-sensory needs. Beginning at least as early as the third century B.C., the more radical Chinese thinkers could openly assert that man, rather than ancestral or nature spirits and the gods, lay at the center of rites and ceremonies. Wearing the ceremonial robe and making the prescribed gestures were outgrowths of the human drive to give form, permanence, and resonant power to feeling. Thus Hsun-tzu, a philosopher who lived in the period 340–245 B.C., opined that sacrificial rites had their origin in longing for the dead and in emotions of remembrance. "Everyone," he wrote, "could be visited at times by sudden feelings of depression and melancholic longings. . . . If these feelings came to him and he was greatly moved, but did nothing to give them expression, he would be frustrated and unfulfilled, and he would sense a deficiency in his behavior."[13]

A properly conducted ritual offered a broad range of satisfactions, according to Hsun-tzu. Everything in it, from food and music to physical setting, catered to a human being's sensory-aesthetic needs. Thus, "grain-fed and grass-fed animals, millet and wheat, properly blended with the five flavors, please the mouth. Odors of pepper, orchid, and other sweet-smelling plants please the nose. Beauties of carving and inlay, embroidery and pattern please the eye. Bells and drums, strings and woodwinds please the ear. Spacious rooms and secluded halls, soft mats, couches, benches, armrests and cushions please the body. Therefore I say that the rites are a means of satisfaction."[14]

Since the rites sprang as much from human feeling as from the powers of heaven and earth, they could change as sentiments and attitudes changed. When, in 629, a counselor to the Emperor T'ai-tsung objected to the site for the annual spring plowing ceremony on the grounds that it was "not in accord with ancient ritual," T'ai-tsung replied, "Rituals arise out of men's feelings, so how can they remain permanent?"[15] Change within tradition seemed entirely reasonable to the emperor. As understood in the early T'ang dynasty, an imperial ritual was a complicated multimodal artistic-religious affair, the beauty of which reflected both human feelings and the order of the universe. The purely religious element, such as it was, lay in the belief that benign energies released in correctly conducted rites could ensure harmony in society and nature.

The ideal Chinese order was deeply moral-aesthetic. Brute force and violence had no place in it. Rituals were performed and the rules and codes of daily life followed because people in high stations and low found a certain appropriateness and satisfaction in them, not because they feared punishment by the gods or their representatives on earth. In actuality, punishment of one kind or another was of course indispensable to the regulation of social life. Even an imperial ritual could not be conducted without the threat of punishment (symbolized by the presence of whip-wielding proctors), for, given its duration and complexity, the authorities would be foolish to expect participants to perform their exacting roles solely for reasons of emotional satisfaction and aesthetic pleasure.[16]

The Medieval European Aesthetic-Moral State

The rituals of Christian kingship, which emerged in the eighth century, were a major source of inspiration for the European aesthetic-moral state. No Christian kingship properly so called had existed before then, for the early church saw Christ as the fulfillment of the kingship of David, and *all* believers—not only the ruler—shared his glory. The Franks in the eighth century revised the idea: the ruler alone received David's royal mantle. He was to be anointed like David, and he alone was to be crowned. What the New Testament promised to all faithful Christians came to rest visibly on only one head. Kingship thus instituted by the Carolingians was taken a step further by their tenth-century successors in Germany, the Ottonians, who saw themselves as God's deputies on earth. Seated in majesty, a European monarch listened to the choir singing the praises of Christ and himself. "Such rites immensely impressed contemporaries; at one of William the Conqueror's crown-wearings, a bystander was so overcome that he cried out, 'Behold, I see God!' "[17]

Anointment and coronation, once established, persisted through the millennium. "Not only were the main features of these rituals directly transmitted from the Carolingian period, but there was continuity in detail: the coronation prayer pronounced over the last of the Bourbons, Charles X, in 1825 was essentially the same prayer as had been first used for his Carolingian ancestor Charles the Bald in 869."[18] In our own time, in 1953, hundreds of millions of people throughout the world witnessed, via television, the coronation of Elizabeth II in Westminster Abbey. The most poignant and historically resonant moment came when Elizabeth, partially hidden under a canopy and dressed in a simple robe, was anointed queen. Good theater, some would say; others, even more dismissively, saw in it only the ermine and glitter of a Hollywood-like extravaganza. Nevertheless, most British and Commonwealth spectators probably thought that they had witnessed something important; they might even have felt a deepening "bond, under God, between sovereign and people." This moral-religious element, inherited from the past, added weight to the visual and aural splendors of the occasion.[19]

In Europe as in China, the moral and artistic seriousness of public rites and ceremonies was repeatedly undermined by the demands and pretensions of secular-political life. Art intended to serve God and to elevate the image of man and society was repeatedly co-opted by rulers to promote their own ambition and gratify their vanity. Not only state visits and the reception of ambassadors, but funerals and religious processions were occasions for the courts to dress up and show off their wealth. "Princes, courtiers, and their ladies went forth in cloth of gold and blazing with jewels. . . . [They] thought they looked 'like angels from heaven,' never suspecting that they imagined angels in their own image."[20]

Worldly pomp, however, was and is not the purpose of state rites and ceremonies; rather, its obtrusive presence indicates a departure from society's original intention, which is to formulate with all the powers of art a credible image of an idealized self, with public good at its core. In the formulation, the moral and the aesthetic are inextricably mixed: the arts, including ceremony and architecture, present and dramatize the good. We have noted how this worked in China. Let us now turn to three cases in the Western world: Renaissance Venice, the France of Louis XIV, and the United States.

Renaissance Venice as
an Aesthetic-Moral State

Venice in the sixteenth century had pretensions to being a moral-aesthetic state. Its rituals, though they might appear as worldly and ostentatious as those of other Italian courts, were steeped in religious values, which in both direct and indirect ways reminded governors and governed of the need to work for social good and justice. As a city built upon the sea, Venice could boast a unique site of great natural beauty. Adding to this natural advantage were its outstanding architectural achievements—preeminently, the vast Piazza San Marco. Both natural and manmade beauty were an unending source of civic pride, not only because they pleased the eye but because to humanists influenced by Neoplatonism outward beauty signified inner virtue. Venetians considered themselves exceptionally pious, and could point to all the tangible signs of piety around them as evidence, including the presence of the body of the Evangelist Mark in their midst, the numerous churches, and the devout processions. In addition, Venice laid claim to such good deeds as patronage of religious orders, charity to the poor, and relentless struggle against infidel Turks.[21]

Renaissance Venice was widely known for the number and grandeur of its public ceremonies; its very name evoked images of pomp and high occasion. Ceremonial display served, however, the serious purpose of presenting in vivid and dramatic form the hierarchical structure of society, which, in the sociopolitical thought of the time, signified not brute domination but the natural order of privileges, duties, and obligations, ultimately underwritten by God.[22] A particularly colorful embodiment of the Venetian social and constitutional ideal was the ducal procession, which began at the palace, moved around the edges of Piazza San Marco, and ended in the basilica. Its course from palace to Saint Mark's tomb linked the authority of the doge and the sociopolitical order of the republic with the hierarchs and emblems of the sacred sphere. Indeed, in a ducal procession the political and liturgical spheres merged even in the smallest details. For example, the white candle carried by the doge's chaplain and regarded as the ruler's personal insignia of privilege and honor was originally a symbol of penitence. The doge played a key role even in the most liturgical feasts. On Palm Sunday he and other magistrates carried gold-leaf palm branches around the piazza in a reenactment of Christ's entry into Jerusalem. And in the dramatization of Christ's last days, the doge himself played the part of Christ.[23]

By the end of the sixteenth century the number of ducal processions had so increased that no less than sixteen could occur in one year. Lavishness also increased, so that the Piazza San Marco had to be periodically enlarged and embellished to accommodate the participants, the spectators, and their rising expectations. Venetian society became more highly structured from the thirteenth to the sixteenth century, and this trend was reflected in the greater importance attached to one's position in the procession. A consequence of this attention to hierarchical order, with participants' relative status marked in striking ways, was spectacles of ever greater magnificence. Such vaunting of class differences did provoke conflicts, both between and within social classes, but these were seldom as severe as those that occurred in rival city-states such as Florence. Among the means of promoting social harmony in Venice, not the least was the granting of a sense of dignity, if not political power, to its citizens through an extraordinary range of locally or professionally generated festivals and ceremonies, culminating in the solemn rites of government, which symbolically embraced the entire community. Besides the ducal processions, the elaborately conducted state ritual known as Marriage of the Sea was of singular importance, for it focused attention on the city's uniqueness, which was a source of good feeling to all Venetians. At a critical point in the ritual, the patriarch emptied a huge ampulla of holy water into the sea, and the doge in turn dropped his gold ring overboard, saying, "We espouse thee, O sea, as a sign of true and perpetual dominion." The Marriage of the Sea and other sumptuous state rites achieved, in Frederic Lane's words, "the artistic mastery of government by pageantry."[24]

The Venetians' view of themselves is what we would now call a myth. A modern meaning of myth is that it is a fanciful story of superheroes, whose doings and achievements bear little relation to the real world. Ritual is the dramatic enactment of myth. Many myths remain purely oral or, as in the case of the American Constitution, textual. Renaissance Venice, unlike the United States, lacked a revered text articulating the formal structures and ideals of society; for Venice, the state rituals were the constitution. As the American Constitution is both "unreal" in its exalted utterances and yet has had a major impact on American society, so the Venetian view of themselves—their enacted myth—was both unreal and real: unreal as all presentations, bracketed off from the routines and makeshifts of daily life, are unreal, but real in its ability to prod society to move closer to an ideal of the good.

What was the social reality? Venice aspired to be a moral state, not just an adornment on the coast of the Adriatic. It believed that it had created a form of government that promoted domestic harmony. Even the poor, Venice boasted, were adequately fed. For centuries it enjoyed a reputation, reaching far beyond its borders, for equitable administration of justice. Nobles and commoners had equal standing in court. Jean Bodin, a prominent contemporary French critic of the myth, nevertheless admitted that in Venice a gentleman, should he do harm to even the least of the city's inhabitants, would be severely punished; and "so a great sweetness and liberty of life is given to all." Prisoners too indigent to hire a lawyer were provided an ad-

vocate from Venice's corps of licensed lawyers.[25] The doge in his oath
of office swore that he would see equal justice done to all, great and
small. The reality may often have fallen short of this ideal, but it
might be argued that there would have been more miscarriages of
justice and that even the idea of justice would have faded if a standard of the good had not been repeatedly and vividly brought to the
people's attention in ceremonies and rites.

France: The Sun
King's Aesthetic-Moral State

In the fifteenth century the humanist Pier Paolo Vergerio expressed
his admiration for Venice by saying that its form of government was
"mixed": Venice was an aristocratic republic that also included some
democratic and some monarchical features. Bodin faulted Venice for
the same reason. As a citizen of monarchical France, he was offended
by the idea of a mixed state. Sovereignty, for him, was by definition
indivisible and could be found only in one place, one institution.[26]

"Mixed" government was aesthetically as well as politically offensive. It suggested conflict, compromise, and makeshift. France
opted for lucid order, which meant monarchical government and a
single source of power, splendor, and virtue, fanning outward to the
remotest reaches of the state. An approximation to this ideal was
first attained during the long reign of Louis XIV (1643–1715), the
Sun King; and Versailles, the seat of court and later (from 1682) of
government, became its architectural embodiment. Sun symbolism
was everywhere. Three straight avenues radiated outward from the
front entrance of the vast palace. Behind it and the formal garden
stretched the "green carpet" axis, which extended to the distant horizon by means of a canal. In 1690 the king's bedroom was shifted to
the center of the palace: the rituals of his rising in the morning and
retirement at night marked the course of the dominant star in
heaven. Many elements in the gardens were placed in relation to the
sun or to its personification.

On the south front, for example, the flower garden and orangerie
emphasize in symbolic decorations Flora and her lover Zephyr,

and Hyacinth and Clytie (heliotrope), turned into flowers by Apollo. The legend of the Sun King and his universe is echoed in the statues of the seasons, of dawn, midday, evening, and night, of the four continents and four elements of earth, air, fire, and water. The Fountain of Apollo, at the end of the canal, shows the sun god in his horse-drawn chariot, emerging from the sea. His return from his daily journey to the ministrations of the nymphs of the sea goddess Thetis is depicted in the Grotto of Apollo's Bath.[27]

Nature beneath the orbit of the moon may have its vagaries, but the majestic course of the sun is dependable. So also were the daily routines of the Sun King. Louis wished his timetable to be so inflexible that anyone in Europe could know at any moment what the king of France was doing. Moreover, as the sun is dazzling, so must be the quarters of the king. A dazzling world was made possible by France's artistic and technological supremacy. Versailles benefited from such outstanding talents as the architect Le Vau, the painter and decorator Le Brun, the landscapist Le Nôtre, and hydraulic engineers at the peak of their newfound powers. In its endless rounds of fêtes and celebrations Versailles showcased numerous artists and their sparkling productions, including Molière and his plays and Lully and his music. There was also an abundance of dazzle in the literal sense. Marble, gold, and silver were everywhere. The palace "boasted furniture of solid silver—tables, consoles, chests, candelabra, and large square planters holding miniature orange trees. Add to that the gleam of the silks and velvets worn by the palace's occupants, and the fire of their jewels," and we have an idea of the special quality of illumination that could be attained long before the age of gaslight and electricity. Over the vast grounds of the estate, flowers fresh from the hothouses and transplanted around the palace at dawn every winter's day presented an image of brightness, as did the sparkle of water shooting skyward from the fountains. And last but not least, the Sun King himself dazzled the eyes of the courtiers—the dazzlement of power made literal by the resplendent suits he wore on grand occasions, including one with more than 12 million livres' worth of diamonds sewn onto it.[28]

All this artistry and show would have been gross vanity without the rationale of a social-moral purpose. And there was one, which Louis took very seriously. He once declared: "We must consider our subjects' good before our own. . . . We must give them laws for their own advantage only; and we must use the power we have over them only so as more effectively to bring them happiness."[29] Throughout his life Louis aspired to glory, but glory meant not so much military prowess as greatness, boldness, and pride, as well as compassion and justice. Glory also meant being seen and sharing. In his own visibility and ease of access Louis contrasted himself with kings and grandees who believed that majesty called for living in secluded magnificence and making the rarest of public appearances. In Louis's view, such isolation promoted terror and servitude, which did not suit the genius of Frenchmen. Louis shared his world; that is, he acted as though Versailles belonged as much to France as to himself. From the start, its state apartments and gardens were open to all decently dressed visitors. By 1685 the gardens so swarmed with Parisians that the king could no longer walk in them without being mobbed.[30]

Etiquette at Versailles was elaborate, as one would expect of a society of subtly and precisely differentiated stations. The distance between a minor nobleman and the king was very great. Whether one stood or sat and, if one sat, whether on a stool or in a chair, were matters of consuming importance. Still, Christianity taught that every person possessed an immortal soul that Christ had come down from heaven to save, and this teaching, with its implied equality, also found expression in etiquette. At Versailles, Louis was famous for never passing a charwoman without taking off his hat to her. Madame de Maintenon considered her sister-in-law vulgar because she sometimes accepted service from a footman without thanking him. Everyone had dignity, which must be recognized. When the Duc de Coislin encountered maidservants in a public place, he would address them as "young ladies" and would put himself at considerable inconvenience to accommodate what he perceived to be their needs. The court expected the highest standard of behavior. Debauchery, drunkenness, vice, and even obscene language were explicitly forbidden.[31]

Versailles—the beauty of the place, its rituals and social dances—was for Louis XIV an idealized model for the whole of France and French society. The rays of the Sun King flooded the palace and its grounds, and, in theory, their power and illumination extended to the borders of the state and beyond. Louis's image of his and France's place in Europe might seem the delusions of an autocrat, yet elements of it were able to take on some substance. Historians assign importance to the great monarch because, in the course of his long reign, his aspirations for himself, for Versailles, and for France were remarkably successful. France's political and cultural leadership was recognized throughout Europe for more than a century. Versailles's standard of beauty and courtesy—how people ideally behaved toward each other in the public realm—remained valid until the Revolution.

America the
Beautiful and the Moral

The leading political and moral idea of our time—democracy—has "a strong aesthetic ingredient," according to George Santayana. Of course democratic ideals were driven not by aesthetic concerns, but by the hatred of oppression and arrogance of rank, and by the aspiration toward a freer and more just society. Nevertheless, "democracy, prized at first as a means to happiness and as an instrument of good government, was acquiring an intrinsic value." It was beginning "to see good in itself, in fact, the only intrinsically right and perfect arrangement. A utilitarian scheme was receiving an aesthetic consecration."[32]

This "perfect arrangement" is moral-aesthetic, and, insofar as they can be separated, the first term is the more important of the two. A purely aesthetic state is unimaginable, for it would be a mere show. The Chinese imperial state, the Venetian city-state, and the French monarchical state all aspired to be good and moral—efficient, uncorrupt, and caring; but also good and glorious, good and splendid, good and harmonious—that is, beautiful. Now we come to the modern democratic state—to America. In what sense is it moral-aesthetic? How does the democratic ideal find expression in architecture, ceremony, and landscape?

Jean Bodin preferred monarchical France's centralized sovereignty to Venice's mixed form of government. What role can the aesthetic fill in a democratic state? The aesthetic implies the existence of a central, organizing intelligence. Common wisdom asserts that "art cannot be produced by a committee." Common wisdom, though, can be wrong. The King James Bible was produced by a committee. Great paintings have been produced by a master-artist and his disciples. Architecture, obviously, calls for collaboration among different talents. Still, in most cases of artistic success some kind of controlling intelligence is at work. Democracy, with its vocation to accommodate multiple—even conflicting—voices, would seem to rule out the possibility of creating anything coherent, much less lucidly harmonious. A further difficulty in reconciling democracy with the aesthetic is that the aesthetic implies excellence, a hierarchy of values, a focal point or several focal points of interest around which lesser ones are arranged. Another hurdle, perhaps unique to the American experience and American democracy, is what the historian Daniel Boorstin calls "vagueness." American space, from the time of the Revolution to the Civil War, was a vast undefined hovering and haunting presence "out there to the West." This vagueness of space had correlates in a certain cloudiness of vision about what the new nation wanted. Boorstin notes that during this period Americans were "distinguished less by what they clearly knew or definitely believed than by their grand and fluid hopes. If other nations had been held together by their common certainties, Americans were being united by a common vagueness and a common effervescence."[33] These fluid hopes and vagueness did not bode well for the creation of beauty, in its prevalent conception as sharply edged and articulated order.

Yet a strong case can be made that America was and is a moral-aesthetic state, that despite the existence of powerful contrary forces and proclivities, attempts have been made to fulfill the conditions considered necessary to achieve that standing. First is the condition of intentionality: the aesthetic is a conscious achievement. We have seen that even in T'ang China, rulers and their advisers sought to use, rather than be bound by, tradition. They were acting in this respect like artists. As for the Italian city-states, Jacob Burckhardt famously called them works of art.[34] And it hardly needs saying that Versailles and its court rituals were a highly conscious artistic creation. Now, America bested these older societies in intentionality because it was from the beginning a deliberative and reflective venture. From the start, it was perceived as a new chapter to be written, a New Jerusalem to be built. "Europe is Egypt and America the promised land"—a message already implicit in Washington's First Inaugural became explicit in Jefferson's Second Inaugural, and henceforth it lingered as a theme in almost all weighty presidential deliverances.

Over the "vague" and "fluid" New World, the Founding Fathers sketched their design of the nation. The design showed strong European influence, even when such influence—in retrospect— seemed inappropriate to a new democratic republic. Hierarchy was recognized and the necessity for protocol and etiquette reaffirmed, though not without controversy. How was the president to be addressed—as "His Most Benign Highness," as John Adams wished for Washington, or simply as "His Highness," as a Senate committee preferred? "Take away thrones and crowns from among men," John Adams wrote in 1790, "and there will be an end of all dominion and justice." Adams hated the plain, managerial style of the president. It lacked all grandeur and could not be expected to command respect from the common people and from foreigners. The Senate debated its own dignity. Should an official statement refer to the "power and splendor" or the "power and respectability" of the new government?[35]

In architecture, splendor triumphed over mere respectability or utilitarianism in the building of the new capital. Chief architect Pierre-Charles L'Enfant, inspired by Versailles, gave the plan a style tailored originally to the pretensions of an absolutist monarch. He conceived the American city as essentially an outdoor palace of radiating avenues and imposing views.[36] Jefferson disapproved of the plan—in particular, of its scale; Washington saw it as a great work of art and gave it his official approval. Whereas Jefferson preferred the new capital to be comparable in size to Williamsburg, L'Enfant planned it on the scale of eighteenth-century Paris: its area of fifty square miles was to accommodate 800,000 people. The architect envisioned plazas and monuments for future glories, including a naval column that would be emblazoned with future victories, and an interdenominational church where the bodies of future national heroes would be interred by act of Congress.[37]

One reason L'Enfant embraced the project with enthusiasm was his belief that no nation "perhaps had ever before the opportunity offered them of deliberately deciding on the spot" the design and construction of a great capital city. Was L'Enfant aware of St. Petersburg, built in the early decades of the eighteenth century? He might have been. St. Petersburg, however, was the project of an autocrat who commanded vast resources, outstandingly labor. Tens of thousands of serfs were conscripted to work on the site. The human toll was horrendous, but construction continued unabated because a supply of labor from the interior seemed inexhaustible. In the end, perhaps as many as 150,000 workers suffered serious injuries or lost their lives. The imperial city rose rapidly out of the mud like a magician's conjuration.[38] Of course, the capital city of a democracy could not be constructed in such a manner. When in 1792 L'Enfant wished to employ a thousand workmen, he was considered unrealistic. Progress crawled. When the government finally moved in in 1800, the city had more mud and empty lots than magnificence. Even in 1842 Charles Dickens, visiting the capital as an honored guest, felt justified in calling it not so much the City of Magnificent Distance as the City of Magnificent Intention.[39]

A large architectural conception, efficiently and rapidly carried out, presupposed a master mind and centralized power that were at odds with democratic sentiment and practice. From the beginning the size and pretension of Washington, D.C., posed a paradox to the young republic. Yet in the course of time the city did achieve the sort of visual grandeur that its original promoters had hoped for. With one part of its soul, American democracy harbored the desire for a certain dignity and grandeur in its public face; and it was its good fortune to have had a number of successes, though not always by the shortest route. The history of the building that houses Congress—the Capitol—is a good example of the convoluted processes of democracy. As befitted a democracy, a competition for the design of the building was held in 1792. (Vestiges of the winning design can still be seen in the rebuilt east facade.) Several major alterations and expansions were undertaken in subsequent decades, preeminent among which were the raising of the great dome and the construction of two outlying wings to balance it in 1851–1867. The changes in design were accomplished in an atmosphere of bitter political and aesthetic bickering. Men freely accused each other of incompetence, extravagance, even venality. Congress and the president were forever interfering. Yet the result is something that the nation can be proud of. The Capitol spoke and speaks to the nation. It inspired a "national" art for later public buildings, including the Department of State, the old General Post Office in Washington, and a number of state capitols.[40]

From its inception as an independent state, America sought to erect buildings rooted in European aesthetics as a part of the image it had of itself. True, by the end of the nineteenth century, America also aspired to create a style uniquely its own—an "architecture for Democracy," as Louis Sullivan put it; but it was never clear what constituted democratic architecture. Would it be buildings that adapted to nature and hugged the soil—the prairie houses of Frank Lloyd Wright, for instance? Yet Wright designed a mile-high skyscraper, and a certain sculptural monumentality looms in some of his most famous works, including the Guggenheim Museum.

If America felt and continues to feel a certain ambivalence toward monumental architecture, it did not feel this way (except for the anxiety in the initial encounter with wilderness) toward the powerful presences of nature. Americans quickly took pride in nature's size and antiquity. The process began with Jefferson's admiration of the vastness and geologic age of the Blue Ridge Mountains, and continued with the subsequent canonization of such geologic and natural wonders as Yellowstone, Yosemite, the Grand Canyon, dinosaur footprints and bones, the giant sequoias and redwoods.[41]

Clearly, there is a disposition even in a democracy to esteem excellence in nature and in art. A hierarchy of values is accepted even though it is in conflict with egalitarian sentiment. But, whereas one can think of L'Enfant as an artist capable of producing artworks, and whereas one can also see God as the creator of nature's wonders, it is less clear how egalitarianism and the many, frequently divergent voices of democratic practice manifest themselves aesthetically. What are some of the characteristic properties of a democratic aesthetic? The words that seem best to describe them are *comely, fair, competent, agreeable, spacious, pleasing, unpretentious, decent*. St. John de Crèvecoeur put it well as early as 1782. In a famous letter he describes what a visitor can see in America. "Here [he] beholds fair cities, substantial villages, extensive fields, an immense country filled with decent houses, good roads, orchards, meadows, and bridges, where, a hundred years ago, all was wild, wooded, and uncultivated! What a train of pleasing ideas this fair spectacle must suggest!" Look at our habitations, Crèvecoeur says; although you will not find grand architecture, you will find "a pleasing uniformity of decent competence."[42]

To Americans, nature can mean the spectacular landmarks that are a source of pride and inspiration. It can also mean open space and free land, a source of spiritual and democratic values, including liberty, simplicity, and equality. George Washington was the first American to link the challenge of frontier space with the renewal of democratic values. European influences, he thought, could be debilitating, but the American people had no reason to fear, because, as he put it in 1788,

> extent of territory and gradual settlement will enable them to maintain something like a war of posts against the invasion of luxury, dissipation, and corruption. For, after the large cities and old establishments on the borders of the Atlantic shall, in the progress of time, have fallen prey to those invaders, the western states will probably long retain their primeval simplicity of manners and incorruptible love of liberty. May we not reasonably expect that, by those manners and this patriotism, uncommon prosperity will be entailed on the civil institutions of the American world?[43]

A century later Frederick Jackson Turner was to take up this idea, develop it, and make it famous. The frontier carried a moral tone, but also a glamor unique to America. Open space has its own exhilarating, magical beauty. When fused with the ideals of simple manners and love of liberty, it turns into a sort of geographic-aesthetic-moral icon.

If egalitarianism can bring about "a pleasing uniformity of decent competence," what of that other democratic value—a government of the people by the people even though they speak with different and unavoidably conflicting voices? Here the answer takes the form of asserting that democracy can somehow achieve an overarching, fluid unity made up of, and sustained by, competing-cooperating groups. A union of diverse elements is a general accomplishment of art. What distinguishes the democratic nation-state as art is the provisional and ever-shifting nature of the consensus and the fragmented, sometimes raucously divisive processes that lead to it. The miracle is how things come together at all. How do thirteen sovereign states, each with its zealous patriots, become the United States? How does a "promiscuous breed" (as Crèvecoeur put it) of "English, Scotch, Irish, French, Dutch, Germans, and Swedes" become Americans?

The fear of the breakdown of the union has haunted the American dream since the founding of the Republic, and it continues to cause anxiety—an anxiety not unlike that of a great artist who recognizes the virtue of richness, who welcomes the clash of colors on the palette, but wonders whether their application on the canvas will result in a coherent composition. In the social sphere, America has successfully created a number of civic rites that enact the process by means of which heterogeneity turns into union, disparate parts into a whole. Preeminent among them are the Memorial Day ceremonies that commemorate the dead. Lloyd Warner believes that they have had the effect of integrating "the various faiths and national and class groups into a sacred unity." Warner sees four stages in the process. In the first stage, the ceremonies actually emphasize separateness: different ethnic and religious groups enact their own ceremonies at different times and in different places. In the second stage, the ceremonies remain separate, but "they are felt to be within the bounds of the general community organization." In the third stage, "there are still separate ceremonies but the time during which they are held is the same." And finally the ceremonies "become one in time and one in space. The representatives of all groups are unified into one procession. Thereby, organizational diversity is symbolically integrated into a unified whole."[44]

Warner adds that the structured process leading from heterogeneity to union is not necessarily known to those who participate. It is, rather, felt by them.[45] A creative force is at work in which the cooperative effort and the principal stages leading to a final unitary goal are not consciously planned. From the standpoint of people's participation and involvement, this process can be described as "from the ground up." Individuals and separate groups move, as though under divine guidance, to a common end. Democracy sometimes works that way. Such, at least, is America's hope. We see it expressed emblematically on the dollar bill as a pyramid rising from its broad base to the as-yet-unfinished apex under the eye of God. This movement from the bottom up is one manifestation of the genius of American democracy. The other proceeds in the opposite direction, from the top down—that is, first the general principles are articulated with maximum clarity, then differences, divergences, subtractions, and additions are allowed to emerge. In the process the initial conception is subtly altered, without, however, losing its original overall shape and intent.

This second, "top-down" process is magnificently exemplified by the township-and-range survey, which Boorstin calls "one of the largest monuments of *a priorism* in all human history."[46] Its impact on landscape and life north of the Ohio and west of the Mississippi remains conspicuous, despite innumerable modifications and adjustments. Yet the Constitution, which presides over the form and operation of American government and hence affects nearly all areas of public life, must be considered an even more important example. Both the survey and the Constitution exhibit the intellectual-aesthetic virtue of clarity. But whereas the clarity of the land survey is simple and geometric, that of the Constitution is complex, one of its principal tasks being to state, as succinctly and precisely as possible, the checks and balances of power.

Americans are often perceived as a practical people, who hold words suspect but excel in doing and making. Yet numbered among their greatest monuments are literary texts such as, besides the Constitution, the Declaration of Independence and the Gettysburg Address. In America, *words* soar rather than temples, shrines, and statues. The architectural historians John Burchard and Albert Bush-Brown admire the Lincoln Memorial in Washington, D.C. They admire, first of all, the seated figure of Lincoln; they like, secondly, the Greek temple that houses the imposing sculpture; and they find much that is right in the building's setting—the great axis that leads from the Capitol to the Washington obelisk and then past a long reflecting pool to the memorial at the bank of the Potomac. And yet they believe that the memorial may owe most of its "final quality" to the text of the Gettysburg Address itself, engraved in one of the halls.[47] What is true of the memorial is also true of the United States on a continental scale: it may be that the nation's natural endowment in plains and mountains and its cultural achievement in farms and cities also owe something of their "final quality" to the hovering and haunting presence of a small number of texts, written not by professional bards but by people actively involved in the governance of the nation.

The Constitution was the scaffold on which America built its claim to being a moral-aesthetic state. Unlike the unwritten British constitution, which was the product of the slow and gradual accretion of customs and traditions, the American one emerged at a particular time. And although a "committee" created it and although the process of creation was fraught with difficulties, the end result has come to be seen as a supremely crafted political document. Its excellence was recognized early. Washington called it "a new phenomenon in the political and moral world, and an astonishing victory gained by enlightened reason over brute force." At the time of the framing of the Constitution and immediately thereafter, the Founding Fathers sensed that history had entered a critical turning point. Washington wrote to an Irish patriot: "You will permit me to say that a greater drama is now acting on this theater than has heretofore been brought on the American stage, or any other in the world. We exhibit at present the novel and astonishing spectacle of a whole people deliberating calmly on what form of government will be most conducive to their happiness."[48]

The aesthetics of the Constitution and of American democracy rests on two principles, well recognized in the history of art. One is unity that encompasses diversity, and the other, perhaps more time-bound to the extent that it is a romantic creed, the belief that a theme can grow, acquiring branches and subthemes, and yet retain its original character. The first principle was defended by James Madison in 1788. "In a single republic," he wrote, "all the power surrendered by the people is submitted to the administration of a single government." The danger of usurpations of power is guarded against "by a division of government into distinct and separate departments . . . and then the portion [of power] allotted to each subdivided [further] among distinct and separate departments." Madison argued for the merit of size and diversity. The state of Rhode Island, left to itself, might be tempted, under the threat of factious majorities, to abandon the popular form of government in favor of an oppressive form of rule independent of the voice of the people. But Madison said, "in the extended republic of the United States, and among the great variety of interests, parties, and sects which it embraces, a coalition of a majority of the whole society could seldom take place on any other principles than those of justice and the general good."[49]

The second principle of art, romantic in its stress on openness and growth rather than on achieved perfection, is illustrated by the subsequent amendments to and interpretations of the Constitution's original articles. Changes occur, yet the spirit remains the same. Franklin Delano Roosevelt, in his First Inaugural on March 4, 1933, was forced by the economic crisis to recommend drastic measures. He sought, however, to reassure the people by saying,

> Action in this image and to this end is feasible under the form of government which we have inherited from our ancestors. Our Constitution is so simple and practical that it is possible always to

meet extraordinary needs by changes in emphasis and arrangement without loss of essential form. That is why our constitutional system has proved itself the most superbly enduring political mechanism the modern world has produced. It has met every
stress of vast expansion of territory, of foreign wars, of bitter internal strife, of world relations.[50]

Modern democracy, then, is antithetical to the aesthetic because
it is—or can seem to be—nebulous, protean, egalitarian, partial to
the average and hostile to the exceptional. But although it may be
disposed toward these attributes, democracy is not confined to
them. True, Benjamin Franklin once described America as a land of
"happy mediocrity"; yet the desire for excellence has always been
there, manifest in its architectural and technological ambitions and
in a childlike admiration for the monumental in nature. Vagueness
may indeed have been a characteristic of American space to pioneers
and the Founders of the Republic; yet this very vagueness has called
forth a compensating desire for lucidity in the lay of the land and in
the articulation of political principles. Egalitarianism, as the slumber of sameness, is indeed anti-aesthetic, but as a moral stance it has
its own soul-stirring power. Moreover, embedded in the egalitarian
ideal is a modesty that may find sensory expression and inspire artifacts and artworks aptly described as pleasing and honest. The protean is contrary to a static notion of the aesthetic, but the aesthetic
is unchanging perfection only from a narrow classical point of
view. From a romantic angle, it easily accommodates a sense of the
provisional, of fleetingness, of growth and change. American democracy contains all these different, even contradictory, elements.
Few citizens can embrace them all. Those who do—the true lovers of
American democracy—must have an appetite for plenitude and
contrarieties—for order and disorder, the grand and the comely, an
overarching unity of purpose and, within it, teeming voices that offer simultaneously division and new life.

V

Aesthetics and Morality

CHAPTER 10

Good and Beautiful

*T*his book begins with the senses and culminates in that great moral-aesthetic achievement, the state. In descriptions of movement, touch, smell, hearing, and sight as sources of delight, as ways of coming to life under the guiding pressure of culture, the question of morality seldom arises. When attention is on the senses, it is easy enough to forget the cultural institutions, social arrangements, and material base that affect them and give them scope and vividness. But when we shift our attention to artifacts—to paintings, sculptures, landscape gardens, buildings, and cities—it is no longer so easy to repress moral questions. The artifacts are wonderful, but at what expense to nature? What are the human and social costs? And when our attention turns to that great artifact the state, moral issues become preeminent in our consciousness; we may even feel that the moral and aesthetic spheres are incompatible, that magnificent buildings and ceremonies are almost a sure sign of injustice in the social realm, that conspicuous public art, by its commanding presence, has served historically to distract people from comprehending their own oppressed condition.

Are the two terms *moral* and *aesthetic*, which I have often hyphenated as though they constitute a continuum of related ideas, contradictory in their essential meaning, however compatible they may be at a superficial level? If the contradiction is real and fundamental, perhaps we can begin to understand it only when we move beyond the terms *moral and aesthetic, good and beautiful*, to the larger one of culture? The question then becomes, Does an irresolvable contradiction lie at the core of human culture? And since culture is an embodiment of human capability, the ultimate question arises, Is human nature fundamentally flawed?

Assumptions
of Compatibility

An intense awareness of conflict between "good" and "beautiful" is a modern dis-ease. In premodern times, their compatibility tended to be taken for granted. Common sense suggests that all societies, to be viable, must entertain some notion of the good. What constitutes the good depends, of course, on a people's particular experiences and values, but, at a general level, good is simply a fusion of moral and aesthetic conceptions, which include a sense of rightness and appropriateness, care and accomplishment, delight in the way things are done and in the things (both natural and artifactitious) themselves.

This common assumption of compatibility has prevailed in widely different places and times. Consider the Zuñi and the Navaho of the American Southwest. A researcher asks Zuñi and Navaho subjects to do two sets of drawings, one "pretty" and the other "ugly." The Zuñi "pretty" drawings emphasize not merely abundance and well-being, but these conditions as the result of labor. The effort itself, skillfully carried out, is commendable—virtuous and pleasing. For the Zuñi, "ugly" scenes show "the difficulties inherent in providing food and housing and the carelessness and maliciousness of human nature." *Ugly* is both a moral and an aesthetic term. It connotes not only human carelessness and maliciousness but also disorder in nature and the difficulty of making a living—of producing a harmonious and nurturing human world. For the Navaho, *pretty* means "a vision of green and summery landscape able to support its animal and human life." The aesthetic force of "green" and "summery" lies in their association with life. To see a green and summery landscape is to feel alive, to have one's senses refreshed, and to know that nature is fertile: any attempt to separate "green" from fertility, the aesthetic from the primary impulses of living, is arbitrary and serves only to diminish both. "Ugly" drawings, by contrast, frequently show "the disruption of the natural order—enduring hardship, arid land, illness, accident, and aliens." "Ugly" is a dysfunctional body (illness), nature (arid land), and society (aliens).[1]

In traditional China, the aesthetic was always intimately entwined with other values, including the life-force and morality. Consider landscape. The imperial park of the early dynasties—an ancestor of the garden and of landscape painting—was never intended merely to provide a view. Rather it was a world of natural and supernatural forces and spirits, a world designed to assuage the human longing for immortality. These mystical-magical attributes of landscape later became diluted, but they never totally disappeared. Primitive forces of chthonian religion, shamanism, and magical Taoism yielded in time to the refined spirits of philosophical Taoism, enlightened Buddhism, and discriminating Confucianism. Grotesquely shaped rocks, suggestive of primordial monsters, remain in the eighteenth-century mandarin garden, but its overall atmosphere is serene and the mood it evokes contemplative. In the garden one enters a world far removed from the human vanity and strife of social life at all levels, from home and marketplace to court. The garden (and likewise the landscape painting, with which it is closely linked) is thus a moral reality, not only an aesthetic one. The bond between the aesthetic and the moral is even more evident in that view of the cosmos based on the principles of *yin* and *yang* and of the cardinal points. Words used to describe the cosmos, such as *harmony*, *balance*, *propriety*, and *rightness*, indicate the nature of the bond. We can perhaps capture the moral-aesthetic flavor of the Chinese through two caricatures—one of a Confucian, the other of a Taoist. A Confucian is so imbued with the idea of rectitude that everything he sees and does is judged by it—whether it is the names assigned to things, the shape of meat in a dish, the color of a ceremonial robe, or the plan of a city. A Taoist is someone so imbued with the idea of *tao* in nature that everything he sees and does is judged by it—whether it is how he chops wood, how he positions his hut in the mountains, or his moral posture toward society.

In the Western world the link between the beautiful and the good is, historically, just as intimate. The link is made up of many interwoven ideas, difficult (and unsatisfactory) to disentangle. One idea is that of the "whole": the whole is complete, without defects, healthy and holy. Whole, healthy, hale, and holy are etymologically related. Moreover, they may be linked to the Greek term *koilu*, "beautiful." Whether or not such a link exists, it is certainly the case that whole and healthy beings are pleasing to the eye. They are life-enhancing—aesthetic. To the Hebrews, sacrificial animals must be whole, pleasing to God; eunuchs, physically incomplete, could not be priests. To the Greeks of classical antiquity, a beautiful body implied a beautiful soul; gymnastics was not mere body-building. Plato puts forward the influential idea that beauty, accessible to the senses, is the only spiritual thing that we love immediately by nature; he treats beauty as an introductory section of the good. Beauty and morals (that aspiration toward the good) are to be seen as parts of the same structure.

The association of outward beauty with inward grace (the body as a temple of the soul) persists throughout the history of the West. Christ is never described in the Bible, yet Western artists have again and again portrayed him as physically attractive, despite the disfigurations of suffering on the cross. The earliest bas-relief depictions of Christ show him as a handsome youth.[2] The great Renaissance masters, da Vinci and Michelangelo, made Christ beautiful. In literary works, too, good people are often also comely people, even erotically appealing. Think of the handsome peasants in Tolstoy—

in particular, the serf boy (with his row of shining white teeth) who so selflessly and unself-consciously serves the bodily needs of the dying Ivan Ilych. Think of Billy Budd, the handsome sailor and a Christ figure. Billy is a "peacemaker," whose presence alone transforms the crew's quarters from "a rat-pit of quarrels" to a haven of mutual regard. As the merchant captain puts it, "a virtue [goes] out of him, sugaring the sour ones."[3]

Not only people but also the physical environment can combine beauty with virtue. The Greeks associated wholeness and health with environment. The island of Cos was a great center of health, presided over by Asclepius and his temple; but even more it was the place itself—the sanatorium, the cloistered calm, the noble landscape freed from the clutter and noise of the city—that made for health.[4] Mental well-being also benefits from an orderly setting. Plato's Academy (ca. 387 B.C.) was located in a garden. The monastic cloister—also known as paradise—encouraged contemplation and learning. In America, many seminaries and colleges were originally designed to be oases of calm beauty in the countryside. Planners assumed that these secluded campuses possessed a virtue that could be passed on to the students.[5]

Toward wild nature, the Hebraic-Christian tradition showed a deep ambivalence. Wilderness was darkness and chaos, the realm of Satan and his minions. On the other hand, the barren desert and mountain were also places where God could have direct encounters with his people in a state of nuptial bliss (Hosea 2:14–17). John Cassian (360–435), who exerted considerable influence on Saints Benedict and Gregory, and hence on early Christian monasticism, envisaged the desert hermits as enjoying "the freedom of the vast wilderness" and "a life that can only be compared to the bliss of the angels." Natural landscapes were somehow "pure" and "free." Moral qualities accrued to them. This positive outlook on nature gained increasing acceptance as settlements multiplied and the fear of the unknown diminished during the late Middle Ages.[6] The Cistercians ventured into the forests to establish their monasteries. Their order enjoyed a powerful resurgence under the leadership of Saint Bernard (1090–1153), who was among the earliest European thinkers to appreciate the beauty of the natural environment and attribute to it moral quality and power. The site of the Abbey of Clairvaux he described as having "much charm"; moreover, it "greatly soothes weary minds, relieves anxieties and cares, and helps souls who seek the Lord greatly to devotion, and recalls to them the thought of heavenly sweetness toward which they aspire."[7]

To Saint Francis of Assisi, the earth and its creatures had beauty and goodness because God created them. This attitude, unusual in his own time, became increasingly acceptable with the rising popularity of natural theology, which, unlike the more traditional theology based on a detailed exegesis of the holy book, saw God as creator-artist and providence. God was revealed in nature. Shakespeare found "sermons in stone." Everywhere one looked, one should be able to see the wisdom and beauty of God. Deism built on this theological foundation, as did Romanticism later. Whereas deists stressed nature's rationality, Romantics stressed its splendor and loveliness, inspiring poets and painters, with results of varying quality, from the late eighteenth century onward. The American people were strongly affected. In the nineteenth century, they came to see towering peaks and serrated ranks of lofty trees, suffused by morning or afternoon light, as nature's cathedral to be approached and entered in veneration. Nature was not only sublimely beautiful but also a moral presence, which Americans neglected and abused at the risk of their souls. This attitude remains potent today: contrast the grandeur and purity of the American West as photographed by Ansel Adams with the tourist camps at Yosemite, packed with people and their trash, and the word *desecration* inevitably comes to mind. Since the 1960s the ecological movement has reinforced this American proclivity to see not only beauty but also moral elevation in nature.

The Artisan-Artist
and the Aesthetic-Moral

We are all cultural beings, makers of culture, artisan-artists. To make anything at all calls for attention, imagination, and skill.

These qualities cannot be had without discipline, the effort to over-come self-indulgence of body and mind to add something tangible to the external world. In the case of a large project—a sculpture in stone, for instance—physical effort, sweat and strain, is required, and we can imagine the artist putting his own health at risk to ac-complish a cultural end. Less visible is the mental effort, which de-mands moment-by-moment choice. Such choices almost always have a moral component, as Malcolm Cowley indicates from the viewpoint of a writer. There is a sense, he says, in which not only fic-tion, "but all kinds of writing, are moral. Almost every work set on paper involves a choice. The writing of one page might involve hun-dreds of moral decisions. 'Shall I use this word, which easily comes to hand, or shall I stop and search my mind for a slightly better word?' " These aesthetic decisions take on a moral aspect if only be-cause they call for "choosing the hard over the easy."[8]

Artifacts of all kinds, including ordinary tools and utensils, can be things of beauty, valued possessions of homes and museums. Artisans themselves have traditionally striven to make objects "good"—well put together, without showiness—rather than mere-ly functional. The best representatives of this genre undoubtedly have an aesthetic appeal. One says of these pots and spinning wheels, these quilts and benches, that they are sound, honest, handsome—that they have character. The words have a moral tone to them. Purely aesthetic epithets such as "elegant" and "lovely" seem inap-propriate, not so much because a bench (say) is an ordinary use-object, but because it signifies a serious way of life. A simple bench speaks of toil and communality: it calls to mind a craftsman's gnarled hands and peasants dozing or gossiping in the sun. The fact that the bench has been used in a certain way, by people who have worn the surface smooth over the years, gives it a look of import—even of holiness.[9]

What is true of utensils and implements is also true of the larger artifacts and landscapes—cottages and barns, plowed fields and ter-raced paddies. The countryside, even more than the particular arti-facts in it, evokes nature and all the warm feelings that nature com-mands. Peasant farmers are rarely seen as working *against* nature, altering it by force. They belong there, and what they do is an artful extension of nature's own essence. This illusion holds even in the case of terraced fields, which can be monumentally sculptural. Most people see beauty in a well-cared-for countryside; they may also agree that its beauty is inseparable from a certain moral ambience and weight.

The Prestige and
Authority of Art

Art is not mere playfulness. When Michelangelo's *Pietà* was slightly damaged by a madman in 1972, the event made the front page of newspapers not only in Europe and North America, but in India: a slight chip on the marble brow of Mary was able to distract the world's attention from the daily maiming and slaughter of human beings in Vietnam at the height of a brutal war. This is one type of evidence of art's prestige. Another type is the systematic and sometimes savage effort to suppress or tame art under totalitarian regimes. Dictators seldom consider even abstract painting a harmless activity that can be left unsupervised. They know art's power well enough to use it as an instrument of enthrallment.

Art can be used in a number of meretricious and evil ways. In itself, however, it is a power for good—a positive, humane and humanizing force. For a start, art can make us happy. "Is it the essence of the artistic way of looking at things that it looks at the world with a happy eye?" Wittgenstein asks, and goes on to say that "there is certainly something in the idea that the end of art is the beautiful. And the beautiful *is* what makes happy."[10] Beauty expands our senses and minds: happiness is living more fully—eyes filled with the loveliness of landscape, ears filled with the sweetness of melody, mind with the eloquence of an argument. A happy person has something to offer—his happiness. "It is very true," says the philosopher Emile Chartier, "that we ought to think of the happiness of others; but it is not often said that the best thing we can do for those who love us is to be happy ourselves." Young people have one thing to offer their elders, which never ceases to impress, namely, their beauty, which is often simply "the bloom of happiness."[11]

In contrast to the flux and muddle of life, art is clarity and enduring presence. In the stream of life, few things are perceived clearly because few things stay put. Every mood or emotion is mixed or diluted by contrary and extraneous elements. The clarity of art—the precise evocation of a mood in the novel, or of summer twilight in a painting—is like waking to a bright landscape after a long night of fitful slumber, or the fragrance of chicken soup after a week of head cold. Clarity, along with vividness, is thus aesthesia—the senses coming to life, which makes for the sort of happiness described just above. Besides the virtue of clarity, art makes transient subjective feeling objective and enduring. One can return to summer twilight in a picture again and again. One can, moreover, communicate how one feels to another by pointing to a picture or a poem. Art is thus a means of establishing mutuality—a shared world. And because it has a degree of permanence, it is able to bridge the continually widening chasm between past and present. Through art, W. H. Auden says, we "are able to break bread with the dead, and without communion with the dead a fully human life is impossible."[12]

Art persuades us to attend to the ephemeral and the insignificant—to events, things, and people that pass us by as we pursue goals that seem to have permanence and that loom large in our perceptual fields. Artists teach that certain things and events have significance precisely because they do not last. True, we are disappointed when soap bubbles break too quickly, but Robert Grudin reminds us that we "would be equally deprived of joy if we could keep them in the cupboard and whack them around like golf balls. They are pleasing not through shape or duration but through a unique relation between the two. A chord of music, a shared glance, a handshake or a kiss loses significance if it is ended too quickly or held too long."[13] Mortal things, says Freya Stark, can acquire a dignity unattainable by the gods by the mere fact of their mortality. "The gaiety of young men going to battle, the mellowness of crumbling walls, the grace of flowers, the delicacy of age, all fragilities glow as it were in the light of their own annihilation."[14] It is not surprising that artists should want to capture evanescent moments of beauty, but they have been known to do so with materials that themselves do not last. J. M. W. Turner was such an artist. "He used watercolor on top of oil and oil on top of watercolor, he worked soft resin into his paint, he used pigments that he knew would fade or darken. It wasn't that he didn't know any better—he did. But the important thing to him was the moment, to capture those evanescent moments and atmospheric effects, and he was willing to try all sorts of unsound experiments and techniques to that end."[15]

The evanescent moment, to those who truly see, evokes paradoxically a conviction of the eternal. The bloom on the tree and the rising curve of the ballerina's arm, the respective beauty of which depends on its fleeting existence, seems, for that reason, to lie beyond temporality. "Art shows us," says Iris Murdoch, "the only sense in which the permanent and incorruptible is compatible with the transient."[16] Experiencing and knowing this conjunction of incommensurables alters the way we perceive the universe: the things that are easily and usually overlooked acquire, under certain circumstances, the authority of a quasi-mystical presence. Our sense of value, our moral outlook, is altered thereby.[17]

In Western thought under the influence of Plato, and in Indian thought as well, the good and the real are "one."[18] Art—good art—has the power to shake us out of solipsism—this common tendency to weave consoling fantasies around our lives. True, quite ordinary kinds of reality can and often do rudely invade and break up our dreaming consciousness: thus the unexpected shower, the blast of music from the neighbor's window, a police officer who has just stopped our car. But an artwork, even though it too has the power to surprise, does not rudely intrude on our subjectivism. Rather it invites us to shed our egotistical anxieties and enter another world—a better world because more authoritative and real. Iris Murdoch's character Dora Greenfield, in a muddle about the direction of her life, went into the National Gallery to pass the time. She looked at the pictures.

She marvelled, with a kind of gratitude, that they were still there, and her heart was filled with love for the pictures, their authority, their marvellous generosity, their splendour. . . . Here was something which her consciousness could not wretchedly devour, and by making it part of her fantasy make it worthless. . . . But the pictures were something real outside herself, which spoke to her kindly and yet in sovereign tones, something superior and good whose presence destroyed the dreary trance-like solipsism of her earlier mood.[19]

The real may not be beautiful or pretty in a conventional sense. Poetry, as Wordsworth says, must not depict only Arcadian bliss. The inhabitants of Grasmere may be picturesque, but they share the human lot of pain. "Is there not an art, a music and a stream of words that shall be . . . the acknowledged voice of life? Shall speak of what is done among the fields, done truly there, or felt, of solid good and real evil . . . ?"[20] The great nineteenth-century novels looked unflinchingly at the depths and heights of the human psyche, the vast array of social types, the degradation of poverty in the midst of affluence, the subtle cruelties within respectable and respected institutions, the messiness and exaltation of life. Reading them broadens our knowledge, enlarges our human sympathies, and forces us to attend to realities beyond the self, which may be radically different from our own, and yet have the power to penetrate and nest in the core of our being. As we read, and for a time after the book has been laid aside, we may feel refreshed and instructed—morally instructed.

A common opinion has it that artists must not preach. "Nonsense," replies the novelist Joyce Cary. "All serious artists preach—they are perfectly convinced of the truth as they see it, and they write to communicate that truth." The preaching—the moral message—may lie in the sensibility of the presentation. "We are struck, for instance, by a certain dignity and distinction in [Tolstoy's handling of 'The Kreutzer Sonata']. With all its violence it has nothing exaggerated, nothing of what we call a false note, and these are moral attributes derived from a moral taste." Written art preaches even when it appears to be merely stating "facts." A novel reveals the moral meaning of the facts—the real, which we can check by reference to our own knowledge of that real. "For example, Dickens's indignant picture of a London waif in *Dombey and Son* made his readers say, 'That's how the poor wretch would feel, would react,' and gave them an indignation that was now also discovered to them as belonging to their own general feeling of what is right and wrong. They had no doubt that their indignation was not merely the consequence of Dickens' skill in working up their feelings, but that it was justified in the actual world. It belonged to their moral conviction about that world even if they had not noticed it before."[21]

Pictorial art, in distinction from literary art, is by its nature less concerned with, and less able to portray, the minutiae of human exchange, the crosscurrents in human relations, that are the daily fare of the moral universe. Pure aesthetic values seem to dominate there; and, as we have noted, people tend to identify the aesthetic with the visual—with color, shape, and composition, with the blending of light green and yellow rose in a kimono sleeve, the foursquare rectitude of a Finnish coffee table, the panoramic view. But even here, moral tones emerge. Light green and yellow rose suggest a certain coolness of self-image—a certain distancing in human relations; the word *honest* comes unbidden as one looks at Scandinavian furniture; and the sweeping view strongly evokes, in some, a sense of liberation. In the history of Western art, moreover, the Christian motif— the life, death, and resurrection of Christ—has made it acceptable to depict not only the tender moments of mother love, but also suffering and torture, the body twisted in agony and stained with blood, the corpse and the death shroud. Stained-glass windows are, to us, art. But they were originally designed to teach the story of salvation to a nonliterate public. Nonreligious art, intended for sophisticated viewers, could also contain a strong didactic element, such as those seventeenth-century still lifes that showed fruits in decay as a reminder of human mortality, and, in our own time, Picasso's antiwar painting *Guernica*.

Pictorial art, again outstandingly in the West, forces us to face certain aspects of reality; Rembrandt's grim *Carcasses*, for example, or *The Good Samaritan*, which shows a defecating dog in the foreground. Over the centuries, people have objected to that shocking detail. Yet that detail, "by rubbing our noses into the world's indifference to good acts, lends power, outrage, and ethical force to the gentle story."[22] The realism of the English painter Lucian Freud, Mark Stevens writes, "is not a matter of the mirror . . . or a clinical examination of the less appealing facts of the flesh. It has a sharper point. Freud's realism has the intensity of sacrifice. Qualms, conventions, hopes—the charms of blindness—must suffer for the sake of truth. The theatrical is everywhere in Freud's art, but as something rejected instead of rendered. He is the master of the unaverted gaze, of the eye that will not blink." In the *Naked Girl* (1966), the subject, believed to be the artist's own daughter, was flagrantly naked, not nude. The figure looked painfully vulnerable. "The eye wanted to recoil. When it did not, however, its reward was power, beauty, and pathos. And one other thing—tenderness. Disgust mastered often leaves behind a feeling of tenderness toward the world."[23]

Though unable to give us the moment-by-moment minutiae of human exchange in a constantly shifting context, pictorial art *is* able to provide not only visual details but also a synchronous grasp of reality—the birth of Christ in one corner of the composition, boys skating on ice, totally indifferent to the event, in another corner—that lies beyond the capacity of narrative art. In their different ways, both stories and pictures present specific and concrete images of life experience.

Music, by contrast, is, in Claude Lévi-Strauss's words, the *mystère suprême* of life. We may not be able to say what it is, but we do not lack words to say what it can do. It can "affect our mental and physical status, the minutely meshed strands of mood and bodily stance that, at any given point, defines identity. Music can brace or make drowsy; it can incite or calm. It can move to tears or, mysteriously, spark laughter or, more mysteriously still, cause us to smile. . . ."[24] For Pythagoras music can heal; for Plato it can madden. To the medieval rhetorician Jean Molinet, "music is the resonance of the heavens, the voice of the angels, the hope of the air, . . . the recreation of all gloomy and despairing hearts, the persecution and driving away of the devils." To Pierre d'Ailly, music's power "is such that it withdraws the soul from other passions and from cares, nay, from itself."[25] In gratitude, Wittgenstein cites Mozart and Beethoven as "the true sons of God."[26] To the writer Gerald Brenan, music offers us "a foretaste of . . . celestial Utopia."[27] The slow movement in Brahms's Third Quartet pulled Wittgenstein from the brink of self-destruction; the same composer's *Alto Rhapsody* seems to have had the same lifesaving effect on William Styron.[28] Joyce Cary writes: "A friend of mine tells me that a Beethoven symphony can solve for him a problem of conduct. I've no doubt that it does so simply by giving him a sense of the tragedy and the greatness of human destiny, which makes his personal anxieties seem small."[29]

These testimonies from different cultures and persons, past and present, suggest that the good and the beautiful, the moral and the aesthetic, are inextricably intertwined—doublets, deeply rooted in common human experience and yearning. Yet, especially in the West and especially since the nineteenth century, a rivalry, even antithesis, has developed between aesthetics and morality. Even a book whose primary purpose is to redirect our attention to the wonders of the human senses, mind, and cultural-aesthetic achievements cannot altogether escape the shadowy side of the same achievements. The shadows raise moral issues that demand notice. In many instances, the deficiencies and horrors in cultural-aesthetic worlds are consequences of inherent limitations in human culture and nature. Let us confront these shadows—some gray, others pitch-black—and see whether, in the end, we might still judge the human venture worthwhile.

Shadows and Light

Nature and culture: the given and its transformations. This must be the underlying theme of any cultural-aesthetic geography—a theme that has been told over and over again in myths, legends, folklore, and now scholarly books. The most familiar story of all in the Western world is the biblical one of how God overcame primordial darkness and watery chaos to create light and land, and then imposed further order on the land by fencing off the Garden of Eden from wilderness. Within God's created world, the Bible offers two images of perfection: in the first book (Genesis) the informal garden, and in the last book (Revelation) the formal glittering city. Similarly, in the origin myths of other cultures, gods and cultural heroes by their mighty acts have introduced order and clarity to a primordial inchoate nature. The progressive implication of such myths, however, is subverted by the fact that the cultural ideal is attained early. Gods and culture heroes did the initial heavy work; thereafter, human duty has been merely one of maintenance. Nevertheless, the developmental thrust of culture is not lost to human consciousness, for it must reappear in one form or another with the education of every child.

From Nature to Culture

The education of a child is a constant reminder that the child is nature, that its body is nature, with urgencies and drives that—like external nature—must be controlled. So hair is combed and the body decorated or clothed. Culture shapes and represses the body, but, unlike external nature, the body can be only superficially modified, not radically transformed. Throughout an individual's life, it remains animal and passionate. Pure and simple happiness, intense pleasure, and the regeneration of life all depend on the natural functions of a healthy body. Because the sensory rewards of the body are great, arguably no less than those of culture, the "making over" of the body by culture is always tinged by ambivalence. Humans, culturally proud, are more or less resigned to leading double-deckered lives, cultural and biological, the one dominant during the day and the other at night.

Children are taught how to behave. Adults present them with a standard of manners and courtesy—a theater of gestures and movements—for emulation. Impulses of natural kindness and generosity are encouraged; anger and violence are discouraged. Manners, backed by moral beliefs, issue in imperatives of dos and don'ts. Transgressions against them produce, in acculturated members, shock—a deep sense of impropriety, and the fear of relapse into chaos and the subhuman. Of course, societies differ greatly in the size of the membership in which good behavior is expected. At one extreme, good behavior is expected only in one's own tribe; at the other extreme (as in Buddhism), it extends to all living things, human and nonhuman. Societies also differ in the importance they attach to social-moral, as distinct from material, achievements. They differ, finally, in the degree to which they see even material achievements in moral terms, as measures of patience, devotion, courage, and breadth of vision.

Radical Transformation: Nature as Food

Humans sustain themselves by transforming nature. Like all animals, they eat—that is, they turn other animals and plants into the substance of their own body. Foods have to be appropriated—animals chased down, killed, and cooked. In simple and complex societies, alike, people may show no sensitivity whatever regarding food preparation and consumption. The orgy of eating among the Mbuti Pygmies, after the bloody killing and dismemberment of an elephant, is not so different from Roman and medieval European orgies of engorgement, next to kitchens where the slaughtering had occurred earlier.[1] In other societies, killing and eating animals has posed moral dilemmas. The Semang of rain forest Malaya cope with the moral issue by extending the code of gentle behavior to their interaction with animals. To mistreat captured beasts, even to laugh at them, is an offense that may be punished by serious illness.[2] Similarly, for the millions of followers of Buddhism, a central tenet of which is compassion toward all forms of life, the problem of blood spilling is solved by a vegetarian diet.

A common theme in the myths of nonliterate cultures is that people and animals could speak to one another in primordial times. The loss of this power signified a fall and caused the alienation of humankind from the rest of nature. Primitive people thus postulate an early state of communality in which bloodshed between them and animals would have seemed abhorrent and unnatural. Literate cultures, too, have myths that implicitly prohibit bloodshed. In the Bible, Adam and Eve, though they enjoy dominion over other animals, can only be vegetarians if the Garden of Eden is to remain innocent. In China, mythical paradises of Taoist inspiration support their human habitants by delicious fruits and nutritious water alone; no hint of blood anywhere.[3] Culture thus offers stories that hide our carnivorous appetite. The history of cooking, too, as we saw earlier, is one of burying the nature of food under artifices of color and fragrance. In China, where the aesthetics of taste is highly developed, cooking is regarded as an artform worthy of the attention of poets and scholar-officials; but when they talk about food, they wax eloquent about wines, teas, and seasonal fruits, rather than about delicately flavored meats. Western connoisseurs of the culinary art are less shy talking about meats, but even they, in their literary effusions, tend to dwell more on the savor of complex sauces, the vintage of wines, the texture of breadcrusts, the freshness of vegetables, and such triumphant but innocent creations as Peach Melba.

Among the complex relations between humans and nature, the most problematical and guilt-ridden is that with animals. People more easily recognize kinship with animals than with any other aspect of nature. Yet animals are killed and eaten, or otherwise brutally exploited. The well-known human ability to compartmentalize knowledge may well be honed on the practice of keeping separate animal as kin and animal as roast. There, sitting at a table sparkling with crystal and sterling silver, the civilized diner must be most on guard against letting the cultural-aesthetic veil slip. The ultimate crudeness is to expatiate on the life trajectories of the foods one eats.

Architecture and
Voluntary Effort

Another major source of unease is the human cost in cultural-aesthetic achievements. The cost to the individual artist is not a moral problem. The story of Beethoven, sacrificing health and sanity in titanic struggles with counterpoint, is inspirational. It does not detract but rather adds to the prestige of his work. In a large architectural undertaking, however, many workers are involved. Unless they participate willingly, the human cost may cast a shadow on the work. In the past, religion has inspired monumental projects for which people from different layers of society have willingly contributed labor, goods, or both. Thus, during the fifth century gigantic sculptural images of Buddha and the Bodhisattvas were carved in the caves of North China. Wall inscriptions proclaim that "all elements of Northern Wei society, the imperial family, the nobility, monks and nuns, and the common people, contributed to the project."[4] The wealthy supplied money and possibly token labor; the common people supplied labor and what money they could. In Europe, it is Lynn White's view that the great cathedrals of the twelfth and thirteenth centuries "are the first vast monuments in all history to be built by free—nay, unionized!—labor."[5] As Chartres rose from the ground, Robert of Torigni reported glowingly that 1,145 men and women, noble and common people, together dedicated all their physical resources and spiritual strength to the task of transporting in hand-drawn carts material for the building of the towers.[6]

Secular buildings, if they had broad symbolic resonance, could inspire similar devotion. In America, this seems to have occurred for the crew that sculpted Mount Rushmore, the crew that built the St. Louis Arch, and the crew that renovated the Statue of Liberty in 1980. In each case, the workmen's esprit de corps and enjoyment in what they did were almost tangible; making these gigantic artworks was fun. One result of this spirit was a remarkable safety record. Planners had calculated, for example, that the St. Louis Arch would cost thirteen lives, but in fact there were no fatalities.[7] Putting up even ordinary buildings has sometimes been a source of deep satisfaction to workers. Their labor and skill, after all, have produced a new landmark, to which they can point with pride. This pride in having erected a new hospital, hotel, or even parking ramp is more aesthetic than moral: the immediate feeling is more, "Look at this wall I helped put up," rather than, "Look, I helped provide a needed social service."[8]

Edifices such as a medieval cathedral and the Statue of Liberty are loved for reasons both moral and aesthetic. However, in premodern times, large-scale constructions almost always entailed tremendous costs in human lives, so that aesthetic values collided violently with moral ones. Another cost was, and remains, the destruction of a previous humanized landscape. Farms and villages, not only wild nature, might have to be destroyed to make way for cities and parks. In ancient China, moralists (including Mencius) complained about the destruction; they believed that the great parks and gardens of the elite were justified, if at all, only if the common people could occasionally enjoy them. In eighteenth-century England, social thinkers such as George Crabbe and William Cobbett, while they could see merit in enclosed fields and scientific agriculture, also recognized their human cost in the obliteration of small farms and villages.[9] And, of course, in our time we cannot help but recall the debate over the morality of urban renewal, which erected gleaming towers on the bulldozed sites of older forms of urban life.

In addition to the hardship of eviction was the hardship of peasants and soldiers conscripted to work on the great engineering and architectural monuments of despotic societies. Laborers in these societies were abundant and easily replaced; their lives, unlike the temples and palaces they built, were transient, quickly forgotten, considered of little worth. In late sixth-century China, when Buddhism inspired an orgy of construction, one ruthless official ordered seventy-two temples built at great cost in human and animal lives. When a monk rebuked him for the suffering and deaths, he replied that posterity would see only the result—the imposing buildings—and would know nothing of the men and oxen that had perished.[10] Versailles was an architectural triumph for seventeenth-century France, built not only as a suitable stage for an absolute monarch but also to project an ideal of civilized society. In a nominally Christian society, however, human lives could not be openly considered of little value. Hence, workmen injured and killed on the job were carted out at night, so that the king's conscience might be spared.[11]

Such grim tales have endless variations, forming a litany of human pain and death in premodern times. The monuments and edifices themselves served to sustain and enhance the power of potentates situated at the apex of their social pyramid and at the center of their cosmos. Awareness of these human costs, added to the cost to the natural world, has long since affected the way people view architectural achievements. One response has been to shun all exhibitions of extravagance, associating it with the arrogance and moral obtuseness of hierarchical societies. Some have embraced plainness and simplicity as an alternative ideal. To a seventeenth-century Puritan, for example, the "pure style of the uncluttered Quaker meetinghouse and the heavy undecorated elegance of Commonwealth silver" had far greater aesthetic-moral appeal than the "gold and glitter of High Church altars," priestly vestments, "the foppish dress and swagger and the gross superfluities of the Court and the aristocracy." And for George Orwell, closer to our own time, "the natural confinement of the working-class terrace and the small suburban house and garden, or village cottage," were morally and aesthetically superior to "the extended spaces and advertised grandeurs of the richer parts of London."[12]

Monumental Projects:
A People's Creativity

Some monuments of the past stand as monuments in the present landscape. They may be viewed with distaste, as we have noted. Generally, however, they are treated with esteem, as part of a nation's heritage; and this was true even in Communist countries. The limelight cast on the horrors of exploitation in the past might have made the palaces and great houses, identified with the exploiting elite, seem repellent. Instead, they were retained and used by the new rulers and their agencies out of necessity, a commendable desire to make use of available space, and the less commendable desire to wrap themselves in the prestige that these buildings could still impart. The official reason for preservation was that these monuments, built by the people, not by the elite, are in fact a tribute to the people's genius. This apparently rote propaganda contains substantial truth. Obviously, it was the people who dug the ditches, leveled the ground, cut the stone blocks and put them up one by one; it was they who skillfully raised the roof beams, cut the window frames, plastered the walls, planted the ornamental trees; and beyond the building site itself there were armies of masons, carpenters, sculptors, silversmiths, clothmakers, and embroiderers, variously skilled and of varying social status, who contributed directly and indirectly to the creative enterprise.

When we consider what a creative enterprise entails, we must also take into account motivation and freedom of choice in means and ends. Artists, in the Romantic conception of the nineteenth century, were thought to be entirely self-driven. But in fact throughout most of European history artisan-artists received commissions, were told what to produce, and worked under the constraints of economic necessity. Workers on a nobleman's estate, organized into teams supervised by foremen, were far less free than individual artisans and artists selling their skills in the marketplace. Yet the threat of force and its actual use, though undoubtedly hovering in the workers' background awareness, could not be a constant brutal presence. Any large construction project that required complex collaboration among numerous specialized and nonspecialized teams presupposed a certain willingness on the part of the workers to exercise their skill and imagination for common ends. This would be true even in routine tasks, and all the more true in tasks that were nonroutine and difficult. In short, Versailles was indeed the work of the people, as it was also the work of Louis Le Vau and Louis XIV. It was an achievement of French society. All capacities for work, all competencies and talents, all powers of coordination had to be in place for it to be built and, three centuries later, still command universal admiration.

State Theater: Images
of Virtue and Splendor

Monumental architecture is frequently a setting for state theater. Graceful gestures, orderly processions, and eloquent speeches in the midst of imposing buildings and spaces are intended to create an idealized model of human worth. They dramatize the distance between people as biological beings, responding impulsively to the urges of nature, and people as civilized and spiritual beings, responding in measured ways to such ideals as harmony, justice, and truth.

Of course, the rituals of state—models of what society should be like—differ, often widely. The traditional Chinese one, for instance, stressed ceremonial gestures and movements rather than architecture. The halls, temples, squares, and avenues of the capital city were essentially the stage, or background, for the performance of the rites. By contrast, in the court of Louis XIV, architecture was as important as ceremonial. The splendid setting of Versailles palace, the hundreds of fountains that were turned on during a state occasion, were central to the theater, not just background. In the United States, although both architecture and ceremony matter, the role of speech is crucial and unique, going beyond a few incantatory words to lengthy invocations of providence, virtue, and the wisdom of the Founding Fathers.

The degree to which nature is incorporated into state theater also varies significantly. In an agricultural civilization such as China's, cosmic forces and the swing of the seasons were thoroughly integrated into the imperial rites: the emperor, as Son of Heaven, mediated between heaven and earth. By contrast, local geography rather than the cosmos played a preeminent role in the state rites of a commercial power like Venice: in that impressive ceremony the Marriage of the Sea, the doge tossed his gold ring into the Adriatic, thus symbolically binding Venice with its milieu—the geographic feature that gives it its singular distinction. Nature took a back seat in the state rites of absolutist France. Versailles itself was a display of human domination: nature was trimmed and dressed, made to behave, as it were, like courtiers. The presence of the Sun King and of the many solar symbols in the landscaped gardens suggested that even the weather must defer to a greater power, that when the Sun King emerged, the clouds dispersed. In American state ceremonies, as distinct from folk rituals like Thanksgiving, nature is curiously neglected. Human power over nature is so much taken for granted in America that it appears to have no need for symbolic enactment. However, as the ecological crisis looms large in the national consciousness, a ritualized gesture or two enjoining people to live more wisely with the natural world may well emerge.

Finally, we encounter differences in the conception of human worth and in the idealization of social ties. Ritual and ceremony, in general, give people a heightened sense of their dignity, as a group and individually. Every participant is indispensable—the child at the tail of a procession no less than the man, laden with honor, at its head. People who merely watch may also gain or regain a sense of worth as they recognize their role, even if it is a humble one, in the great scheme of things. What offends modern democratic sensibility is the unquestioning acceptance of order and hierarchy in older societies. The modern critic who eschews anachronism in appraising architecture—who can see beauty in a Renaissance palace despite its unjust social origins—must fight a visceral prejudice against the idea that the past's hierarchical order could actually be an ideal. Unless we overcome this prejudice, however, we will continue to misrepresent and misjudge the moral-aesthetic aspirations of other times and places.

The past's emphasis on order seems to us excessive. It implies rigidity—a deep fear of change. What we forget is how radically the meaning of chaos has altered. Chaos now means little more than confusion on a large scale; but to the ancients, and even to seventeenth-century Europeans, it meant something more like cosmic anarchy, a dissolution of creation, a reversion to primordial darkness, violence, and flux. In antiquity, fear of cosmic anarchy was so deep and pervasive that extreme measures, including human and animal sacrifice, were considered necessary to prevent it. Early modern Europeans no longer made such sacrifices, but total chaos remained a looming threat: hence their emphasis on order in the social realm to compensate for the unpredictability of the natural realm, and to coax, through sympathetic magic, a greater order in nature itself.

Order need not be hierarchical, but most earlier cultures took it to be so. Thus, to the Chinese, heaven was not just complementary but superior to earth, and both rose above the myriads of local spirits and deities. Human ranks dovetailed with the nonhuman: the emperor himself was a sort of deity who enjoyed a higher status than that of the local spirits of mountain and water. Within the human-social sphere, the reigning model of hierarchy was the family: natural inequality within the family—parents' power over children, and children's dependence on parents—was extended to all relationships of governor and governed in the empire. The Confucian Chinese did not see the model as one of (naked) power; rather, they saw it as exhibiting virtue and benevolence, the reciprocal rules of obligation and duty. To the Chinese, the family model was so obviously pious and right that to depart from it, such as when a parent-magistrate abused his people, was a deep offense not only against a human institution, but also against a universe that had moral order inscribed at its core.[13]

The family model of society—or benign paternalism—found favor in other parts of the world as well, including Europe. But Europe also possessed another model of hierarchical order, cosmic in scope, called the great chain of being. Here is Sir John Fortescue's account from the fifteenth century:

> In this order hot things are in harmony with cold, dry with moist, heavy with light, great with little, high with low. In this order angel is set over angel, rank upon rank in the kingdom of heaven; man is set over man, beast over beast, bird over bird, and fish over fish, on the earth, in the air and in the sea: so that there is no worm that crawls upon the ground, no bird that flies on high, no fish that swims in the depths, which the chain of this order does not bind in most harmonious accord.

Every creature, from the highest angel down to the meanest worm, the writer goes on to say, has a superior and an inferior, "so that there is nothing which the bond of order does not embrace."[14] No creature is superfluous. Mutual need and service give dignity to all.

This conception of hierarchy, with roots in classical antiquity, was still widely accepted as late as the eighteenth century. The minute gradations of rank in the court of Louis XIV, with dress codes and rituals attendant upon them, would have been incomprehensible without the foundational belief in a God-ordained chain of being. And what God had ordained must be good and beautiful.

Shadows: Human
Frailties and Evil

Even if we manage to overcome the profound bias of our time and see a certain moral-aesthetic grandeur in these hierarchical models of society, many doubts remain. They are doubts concerning culture itself, exacerbated and magnified when culture is manifest on the scale and complexity of a political state. I have already mentioned some of the costs to nature and to humans in the processes of making and creating. They are the shadows, inevitable in many instances, if only because destruction of some kind necessarily precedes construction. Let me now add a few more shadows. One is the shadow of stasis or rigidity. If nature is flux, culture is pattern—something sufficiently fixed and distinctive to be consciously sensed and appreciated. Pleasure in an artwork lies in its existence outside the flow of life. It is *there*. But this fixity, in the long run, can be life-destroying. Our feelings risk being imprisoned in the culturally created patterns and objects around us. Ritual carries a potential for deadening formalism; what is intended to elevate life, giving it a magnificence that nature alone cannot offer, ends in tedium. Historical records demonstrate in abundance how the daily imperial rituals of China and the daily royal rituals of Versailles cast a pall over the lives of the privileged—yet simultaneously victimized—courtiers. Almost all human societies exhibit a need to shatter periodically the bonds and boundaries of culture in riotous carnivals, pilgrimages to sacred centers, and bursts of iconoclasm in religion and art; in some societies, the need for a radical break takes the extreme form of revolution and war.

The second shadow is cast, paradoxically, by a human virtue, namely, the power to focus on a project and carry it through with relentless energy. The cost is a distortion of the human world thus constructed. Individual creators may be so consumed by an artistic or research venture that they risk losing their common humanity—their awareness of the needs and rights of other people. In the larger projects of society, similar distortions carry proportionally larger consequences. For example, in the single-minded effort to perfect an imperial rite, in which every piece of the stage, every detail of costume, every gesture strives for perfection, what the rite in totality represents and tries to achieve is easily lost. Conscientious Chinese officials were aware of this danger, and from time to time they creditably urged that the gap between reality and appearance be breached. In the ceremonial life at Versailles, so much energy and attention were given to architecture and to the ceremonial dances of state that their purpose in presenting images and models of ideal human relationships, undergirded ultimately by Christianity, faded into the background. True, they were not obliterated. The splendors of Versailles were not for king and court alone: they were also for the French people, and good manners included thoughtfulness and courtesy to manservants and maids, sometimes carried to ridiculous extremes to illustrate noblesse oblige and the never-quite-forgotten thesis of fundamental equality under God. Nevertheless, outward appearances too easily and too often hid real feelings and relationships; beneath the glitter of candles, the smiles, and the bows lay the beasts of envy, hatred, wicked machinations, even murder.

Shadows are gray if they are cast by such human frailties as inattention, or an excess of attention to the wrong things. They are dark gray if they are cast by flagrant vanity and greed. These darker shadows are common in all societies that have risen above the level of the hunting-gathering band. In these relatively complex societies, all cultural accomplishments that serve to enhance a general sense of human dignity also serve to inflate the vainglorious pride of privileged individuals and groups. The rites of state, with their vivid articulations of rank, tend to sharpen divisions, making members of different groups seem almost different species, rather than creating a community of beings who, though clothed in varying degrees of social worth, are bound by mutual regard and obligation. Furthermore, the artifacts of culture—these embodiments of prestige—are easily appropriated by those who wield them and made to project an aura of legitimacy, however ill-founded.

Finally, there are the pitch-black shadows of evil, of which Hitler's Germany is the most infamous example. Hitler, a failed artist, deliberately used aesthetic culture to create a monstrous state that glorified power for the Aryan race, servitude for "lesser races," and extinction for the Jews. From the invention of the Nazi flag, the swastika armband, and the party uniform, to the mass choreography of the party faithful at Nuremberg in a monumental architectural setting, Hitler harnessed for evil purposes the human tendency to be enthralled by art and show, ritual and ceremony. In contrast to the hierarchical moral-aesthetic states of premodern China, Venice, and France, all of which sought to project an image of benign inclusiveness (whatever the reality in practice), the Nazi German state openly declared its intention to exclude certain peoples even within its own cultural-linguistic boundaries. Moreover, again unlike imperial China, republican Venice, and monarchical France, Nazi Germany made no pretense whatsoever to bestow dignity on its weaker members.[15]

Light: Moral Beauty

Culture is humans' attempt to elevate themselves above nature. However groups may differ, all have a model of the cultured person and bring up their children to conform with it. In the larger view, the human story is one of progressive sensory and mental awareness. If this world seems to us "passing strange and wonderful," it is because culture, through laborious and labyrinthine paths traversed over millennia, has greatly and variedly refined our senses and mind. So when a world citizen now asks, What is a human being? The answer is, Not just a biped animal, but an individual who pauses to smell the sea, listen to silence in the intervals of music, contemplate the shifting spaces of an architectural interior, marvel at what can be seen only with the mind's eye—the curvature of the universe. Hence, from the small child who puts a ringlet of flowers on her head to this world citizen, heir to the wealth of the past, culture bestows dignity. Culture, thus conceived, is a moral-aesthetic venture, to be judged ultimately by its moral beauty.

Moral beauty, narrowly understood, is an attribute discernible in human individuals and in human relationships. A spontaneous act of generosity performed with unselfconscious grace is an example of moral beauty, as are certain acts of courage; genuine modesty is a possible example, as is selfless love. Some people appear to possess moral beauty as others possess physical beauty. Although moral beauty may be a natural gift, it is nevertheless more likely to emerge and flourish in societies that appreciate and encourage it. Certainly, the recognition of moral beauty in another is a more sophisticated accomplishment than the recognition of physical beauty. We admit the difference when we say of moral beauty that it lies below the surface: Socrates is ugly to look at, but he is morally beautiful underneath. "Surface" and "depth" are, however, misleading metaphors, for generous acts, considerateness, wisdom itself, must all be accessible to experience to be real. The quality is there, not hidden, only more difficult to discern.

Relationship with the other is at the heart of morality. Although all societies have conceptions of what this ought to be, the conceptions differ in coherence and comprehensiveness, subtlety and depth—beauty. The scale of a moral system tends to vary with the size of the group. Those of hunting bands are likely to be smaller, less fully developed, than those of large sedentary groups. One of the most admired of these larger moral systems is Buddhism. As the product of a civilization, Buddhism, for all its asceticism and otherworldliness, depends on the existence of an institutional and material base. Footpaths and roads, sacred places, inns for travelers and pilgrims, monasteries are the scaffolds of the Buddhist moral edifice. Material culture, then, plays a necessary role in the invention, elaboration, and maintenance of structures of moral behavior, in the complex exchanges of affection and respect, obligations and duties beyond kin and tribe. Without material and institutional support, moral beauty would remain an individual's accidental endowment, incapable of growth and of permeating society as a whole. Material culture, however, easily aggrandizes at the expense of individual human virtue and at the expense of social relations. It can become monstrous, devouring all talent and resources for its own ends. Buddhist India appears to have avoided this outcome. Confucian China and Christian Europe were less successful: architecture and ceremony there tended over time to be elaborated for their own sakes in disregard of the quality of human relationship they were meant to exhibit and subserve.

America is a much newer experiment in human living, one with moral concerns at its core. In this respect it differs from Europe, which has preferred sophistication and worldly wisdom to "righteousness," and resembles China, which saw the universe itself as essentially a moral order. However materialist Americans may be in their economic pursuits, their ceremonies emphasize the material far less than European societies have. America has imposing official architecture. Washington, D.C., boasts a radial baroque stateliness. Yet one of its most important buildings, the White House, is a modest dwelling, its scale far smaller than that of the palaces of Europe and Asia. The American monuments that inspire open religious awe tend to be works of nature rather than those of man. Reverence for sheer size is indicated in the epithets that Americans have attached to natural monuments: the Giant Sequoia, the mighty Mississippi, the Great Plains, the Grand Tetons, the Grand Canyon. Manmade colossi such as the Empire State Building and the Washington Monument evoke self-admiration rather than religious feeling. The Statue of Liberty may be an exception, but that towering "goddess" uniquely captures the mythic American values of freedom and opportunity.

As with official architecture, so with the rituals of state. In America, they too have a homespun quality. Even the few formal touches they possess are easily modified or discarded, as in Jimmy and Rosalyn Carter's casual substitution of the democratic walk down Pennsylvania Avenue, after the inauguration, in lieu of the traditional motorcade. Popular national rites, such as July Fourth and Memorial Day, tend to be decentralized and informal. Even the Bicentennial was dispersed among many state and local observations rather than centered on one sacred site where a national outpouring of emotion could occur. American rituals undoubtedly have their high and solemn moments, but these are attenuated by the sane feeling that children's laughter, stray balloons, and the smell of popcorn are never far away.

In America, architectural and ceremonial reminders of the dignity of the state and its civic institutions occur everywhere as domed capitols and colonnaded post offices, black-robed judges and smartly uniformed marine guards. In a democratic country, the dignity of the state is that of its citizens. Pride and even occasional pomp have a place in it. But happily, the outward and material forms of dignity are seldom overwhelming. Unlike those of monarchical Europe and imperial China, American state ceremonies do not draw so much attention to themselves that the principles and ideals of human and social relationship, which they supposedly enlarge and embody, fade into the background. And there is another reason why moral arrangements, with their own kind of beauty, do not altogether fade here. Americans are not known to be especially literary, yet it is in this country rather than in European nations and China that words and texts have acquired the public presence of monuments. "We hold these truths to be self-evident, that all men are created equal, that they are endowed by their Creator with certain inalienable Rights, that among these are Life, Liberty and the pursuit of Happiness." These words and others equally well known and revered (sections of the Constitution, the Gettysburg Address, "We shall overcome," "I have a dream") have become a part of the American mindscape. To be an American is to have these words, together with Yosemite, the Empire State Building, and baseball, infiltrate and inform one's consciousness.

There is a significant difference, however, between architecture and ceremonial on the one hand and words and texts on the other. Whereas material symbols, ritual gestures, and music connote, words both connote and denote. The former can become aesthetic objects worthy of admiration in themselves; the latter, for all their artfulness, are also propositions—specific enough to serve as constant reminders of discrepancies between appearance and reality, and hence as agendas for action. Do we have equality? What is our dream? How shall we overcome? If martial music can give us courage, a soaring building make us stand tall, then the weight and moral beauty of words such as those spoken by Lincoln at Gettysburg can heal, elevate, and hint at what it means to be a people.

1. THE AESTHETIC IN LIFE AND CULTURE

1. Montaigne, *Essays*, trans. J. M. Cohen (Harmondsworth: Penguin Books, 1958), p. 250.

2. Helen Vendler, *The Odes of John Keats* (Cambridge, Mass.: Harvard University Press, 1983), pp. 20–39.

3. Paul Goodman, *Five Years* (New York: Vintage Books, 1969), p. 107.

4. Jasper Griffin, *Homer on Life and Death* (Oxford: Clarendon Press, 1983), pp. 20–21, 38–39, 89–90.

5. Carol Z. Stearns and Peter N. Stearns, *Anger: The Struggle for Emotional Control in America's History* (Chicago: University of Chicago Press, 1986).

6. Peter Winch, *Simone Weil: "The Just Balance"* (Cambridge: Cambridge University Press, 1989), p. 114.

7. John Updike, *The Music School: Short Stories* (New York: Vintage Books, 1980), p. 155.

8. Susan Sontag, *Styles of Radical Will* (New York: Dell, n.d.), p. 57.

9. John Osborne, *A Patriot for Me* (London: Faber & Faber, 1965), p. 101; Roger Scruton, *Sexual Desire: A Moral Philosophy of the Erotic* (New York: Free Press, 1986).

10. Roland Barthes, *Roland Barthes* (New York: Hill and Wang, 1977), p. 103.

11. Marshall McLuhan and Harley Parker, *Through the Vanishing Point: Space in Poetry and Painting* (New York: Harper Colophon Books, 1969), p. 75.

12. David Hawkes, ed. and trans., *Ch'u Tz'u: The Songs of the South* (Boston: Beacon Press, 1962), p. 34.

13. Quoted in Henry S. F. Cooper, Jr., "Explorers," *New Yorker*, March 7, 1988, pp. 59–60.

14. Andrew Hodges, *Alan Turing: The Enigma* (New York: Simon and Schuster, 1983), p. 127.

15. S. Chandrasekhar, *Truth and Beauty: Aesthetics and Motivations in Science* (Chicago: University of Chicago Press, 1987), pp. 54, 61.

16. *The Autobiography of Bertrand Russell, 1872–1914* (Toronto: McClelland and Stewart, 1967), pp. 158–159.

17. C. M. H. Clark, *Select Documents in Australian History, 1851–1900* (Sydney: Angus and Robertson, 1955), p. 94.

2. THE DEVELOPMENT OF THE AESTHETIC IMPULSE

1. J. S. Bruner, *Processes of Cognitive Growth: Infancy* (Worcester, Mass.: Clark University Press, 1968), p. 32.

2. "Beauty is in the Eye of the Baby," *Psychology Today*, August 1987, p. 12; report of Judith H. Langlois's research in *Developmental Psychology*, 23 (1987), 363–369.

3. Yi-Fu Tuan, "Children and the Natural Environment," in *Children and the Environment*, ed. Irwin Altman and Joachim F. Wohlwill (New York and London: Plenum Press, 1978), p. 21; see also J. Douglas Porteous, "Childscape," in *Landscapes of the Mind* (Toronto: University of Toronto Press, 1990), pp. 145–173.

4. Robert Grudin, *Time and the Art of Living* (New York: Ticknor & Fields, 1988), pp. 90–91.

5. Richard N. Coe, *When the Grass Was Taller: Autobiography and the Experience of Childhood* (New Haven: Yale University Press, 1984), p. 135.

6. Ibid., p. 210.

7. Vladimir Nabokov, *Speak, Memory* (New York: Putnam's Sons, 1966), p. 24.

8. Pierre Teilhard de Chardin, *The Heart of the Matter*, trans. René Hague (New York and London: Harcourt Brace Jovanovich, 1978), pp. 18–19.

9. C. S. Lewis, *Surprised by Joy: The Shape of My Early Life* (London: Collins, 1959), p. 12.

10. Ibid., p. 11.

11. John Holt, *How Children Learn* (New York: Dell, 1976), p. 76.

12. Keith Swanwick, *Music, Mind, and Education* (London and New York: Routledge, 1988).

13. Ibid., p. 64, citing R. Bunting, *The Common Language of Music, Music in the Secondary School Curriculum*, Working Paper 6 (York, U.K.: Schools Council, York University, 1977).

14. Keith Swanwick and J. Tillman, "The Sequence of Musical Development," *British Journal of Music Education*, 3, no. 3 (1986), 305–339.

15. Swanwick, *Music, Mind, and Education*, pp. 70–83.

16. Rudolf Arnheim, *Art and Visual Perception* (Berkeley: University of California Press, 1956); Miriam Lindstrom, *Children's Art: A Study of Normal Development in Children's Modes of Visualization* (Berkeley: University of California Press, 1974).

17. Lindstrom, *Children's Art*, pp. 48–49; see also Roger Downs and Lynn S. Liben, "The Development of Expertise in Geography: A Cognitive-Developmental Approach to Geographic Education," *Annals of the Association of American Geographers*, 81, no. 2 (1991), 304–327.

18. Howard Gardner, *Art, Mind, and Brain: A Cognitive Approach to Creativity* (New York: Basic Books, 1982), pp. 88–89, 94–95.

19. Ibid., pp. 96–98; S. Honkavaara, "The Psychology of Expression," *British Journal of Psychology Monograph Supplements*, no. 32 (1961), 41–42.

20. Gardner, *Art, Mind, and Brain*, pp. 99–100, 158–167; Kenneth Olwig, "Childhood, Artistic Creation, and the Educated Sense of Place," *Children's Environments Quarterly* 3, no. 2 (1991), 4–18.

3. PLEASURES OF THE PROXIMATE SENSES

1. Toni Bentley, *Winter Season: A Dancer's Journal* (New York: Vintage Books, 1982), p. 138.

2. Roger Bannister, *The Four-Minute Mile* (New York: Dodd, 1955), pp. 11–12.

3. Quoted in Studs Terkel, *Working* (New York: Pantheon, 1974), pp. 385–386.

4. Albert Camus, *Lyrical and Critical*, trans. Philip Thrody (London: Hamilton, 1967), p. 53.

5. Quoted in Philip Hamburger, "All in the Artist's Head," *New Yorker*, June 13, 1977, p. 49.

6. Mihaly Csikszentmihalyi, *Flow: The Psychology of Optimal Experience* (New York: Harper & Row, 1990).

7. Leo Tolstoy, *Anna Karenina*, trans. Rosemary Edmondo (Harmondsworth: Penguin, 1954), p. 271.

8. Quoted in Susan Leigh Foster, *Reading Dancing: Bodies and Subjects in Contemporary American Dance* (Berkeley: University of California Press, 1986), pp. 7, 11.

9. Ibid., p. 16.

10. Ashley Montagu, *Touching: The Human Significance of the Skin* (New York: Harper & Row, 1978), p. 8.

11. Ibid., p. 38.

12. Lisa Heschong, *Thermal Delight in Architecture* (Cambridge, Mass.: MIT Press, 1982), p. 19.

13. Nicholson Baker, "Shoelace," *New Yorker*, March 21, 1988, p. 30.

14. D. H. Lawrence, *Women in Love* (London: Secker, 1921), pp. 120–121.

15. Lorus J. Milne and Margery Milne, *The Senses of Animals and Men* (New York: Atheneum, 1962), p. 18.

16. James J. Gibson, *The Senses Considered as Perceptual Systems* (Boston: Houghton Mifflin, 1966), pp. 100–101.

17. Susanne K. Langer, *Mind: An Essay on Human Feeling* (Baltimore: Johns Hopkins University Press, 1972), vol. 2, p. 259.

18. Bernard Berenson, *Florentine Painters of the Renaissance*, 2d ed. (New York: Putnam's, 1906); see also Ian Chilvers, Harold Osborne, and Dennis Farr, eds., *The Oxford Dictionary of Art* (Oxford: Oxford University Press, 1988), p. 486.

19. Robert Hughes, "When God Was an Englishman," *Time*, March 1, 1976, p. 56; quoted in Montagu, *Touching*, p. 247.

20. Steen Eiler Rasmussen, *Experiencing Architecture* (Cambridge, Mass.: MIT Press, 1964).

21. Simone Weil, *Intimations of Christianity among the Ancient Greeks* (London: Routledge, 1957), p. 190.

22. Simone Weil, *Waiting for God* (New York: Capricorn Books, 1959), pp. 169–170.

23. Colin Thubron, *Behind the Wall: A Journey through China* (London: Heinemann, 1987), pp. 182–184.

24. Bridget Ann Henisch, *Fast and Feast: Food in Medieval Society* (University Park: Pennsylvania State University Press, 1976), pp. 110–111.

25. Elizabeth Burton, *The Early Tudors at Home 1485–1558* (London: Allen Lane, 1976), p. 129.

26. W. H. Lewis, *The Splendid Century: Life in the France of Louis XIV* (New York: Morrow Quill Paperbacks, 1978), pp. 208–209.

27. Robert Mandrou, *Introduction to Modern France, 1500–1640* (New York: Harper Torchbooks, 1977), p. 25. On methods of cooking, see Louis Stouff, *Ravitaillement et alimentation aux XIV et XV siècles* (Paris–La Haye: Mouton, 1970), pp. 258–262.

28. Stephen Mennell, *All Manners of Food: Eating and Taste in England and France from the Middle Ages to the Present* (Oxford: Blackwell, 1985), pp. 146–147.

29. Lady Sydney Morgan, *France in 1829–30* (London: Saunder & Otley, 1831), vol. 2, pp. 416–417; quoted in ibid., p. 147.

30. Jean-François Revel, *Un festin en paroles* (Paris: J. J. Pauvert, 1979), p. 300; quoted in ibid., p. 148.

31. Petronius, "Dinner with Trimalchio," in *The Satyricon*, trans. William Arrowsmith (New York: Mentor Books, 1960), pp. 38–83.

32. Mennell, *All Manners of Food*, p. 160.

33. Georges Auguste Escoffier, *A Guide to Modern Cookery* (1903; reprint, London: Hutchinson, 1957); Mennell, *All Manners of Food*, p. 161.

34. *Li Chi*, trans. James Legge (1885; reprint, Hong Kong: Hong Kong University Press, 1967), vol. 1, pp. 369–370.

35. Quoted in K. C. Chang, ed., *Food in Chinese Culture: Anthropological and Historical Perspectives* (New Haven: Yale University Press, 1977), pp. 37–38.

36. "Lun Yu" in *The Four Books*, trans. James Legge (New York: Paragon Reprint, 1966), p. 130.

37. Arthur Waley, *Yuan Mei: Eighteenth-Century Chinese Poet* (New York: Grove Press, 1957), p. 196.

38. Frederick W. Mote, "Yuan and Ming," in Chang, *Food in Chinese Culture*, p. 238.

39. Ibid., p. 201.

40. Jacques Gernet, *Daily Life in China on the Eve of the Mongol Invasion 1250–1276* (London: George Allen & Unwin, 1962), p. 137.

41. E. N. Anderson, *The Food of China* (New Haven: Yale University Press, 1988), p. 158.

42. Lin Yutang, *My Country and My People* (New York: John Day, 1939), pp. 342–344.

43. Edmond Routnitska, *L'Esthétique en question* (Paris: Presses Universitaires de France, 1977).

44. Sigmund Freud, "Civilisation and Its Discontents," in *The Standard Edition of the Complete Psychological Works of Sigmund Freud*, ed. James Strachey (London, 1975), vol. 21, pp. 99–100.

45. R. W. Moncrieff, *Odour Preferences* (London: Leonard Hill, 1966), p. 270.

46. D. B. Gower, A. Nixon, and A. I. Mallet, "The Significance of Odorous Steroids in Axillary Odour," in *Perfumery: The Psychology and Biology of Fragrance*, ed. Steve Van Toller and George H. Dodd (London and New York: Chapman and Hall, 1988), p. 49.

47. Robert Rivlin and Karen Gravelle, *Deciphering the Senses* (New York: Simon and Schuster, 1984), p. 89.

48. Oliver Sacks, *The Man Who Mistook His Wife for a Hat* (New York: Harper & Row, 1987), p. 159.

49. Berton Roueche, "Annals of Medicine," *New Yorker*, September 1977, p. 97.

50. Sacks, *Man Who Mistook His Wife*, p. 157.

51. T. Eugen, "The Acquisition of Odour Hedonics," in Toller and Dodd, *Perfumery*, p. 80, 85; Moncrieff, *Odour Preferences*, p. 65.

52. Moncrieff, *Odour Preferences*, p. 194.

53. Burton Watson, *Chinese Lyricism: Shih Poetry from the Second to the Twelfth Century* (New York: Columbia University Press, 1971), pp. 42, 87.

54. Alain Corbin, *The Foul and the Fragrant: Odor and the French Social Imagination* (Cambridge, Mass.: Harvard University Press, 1986), pp. 22–23, 32–33.

55. Edward H. Schafer, *The Vermilion Bird: T'ang Images of the South* (Berkeley: University of California Press, 1967), pp. 248–249.

56. J. D. Porteous, "Smellscape," *Progress in Human Geography*, 9, no. 3 (1985), 362.

57. Moncrieff, *Odour Preferences*, pp. 205–206.

58. Leo Tolstoy, "The Hunt," in *Childhood*; quoted in George Steiner, *Tolstoy or Dostoevsky* (New York: Vintage Books, 1961), p. 74.

59. Gerald Brenan, "Village in Andalusia," *The Anchor Review*, no. 1 (1955), 63.

60. Corbin, *The Foul and the Fragrant*, p. 79.

61. "Notes and Comments," *New Yorker*, February 19, 1990, p. 14.

62. Nigel Groom, *Frankincense and Myrrh: A Study of the Arabian Incense Trade* (London and New York: Longman, 1981).

63. Edwin T. Morris, *Fragrance: The Story of Perfume from Cleopatra to Chanel* (New York: Scribner's, 1984), pp. 71, 96.

64. Ibid., pp. 57, 108.

65. Jacques Gernet, *Daily Life in China on the Eve of the Mongol Invasion 1250–1276* (London: George Allen & Unwin, 1962), pp. 120–121.

66. Henry Alabaster, *The Wheel of the Law: Buddhism Illustrated from Siamese Sources* (London: Trübner, 1981), p. 294.

67. *The Meaning of the Glorious Koran*, trans. M. M. Pickthall (New York: Mentor Books, 1953).

68. Colleen McDannell and Bernhard Lang, *Heaven: A History* (New Haven: Yale University Press, 1988), pp. 70–72.

69. Henry Inn and S. C. Lee, *Chinese House and Gardens* (New York: Hastings House, 1950), p. 25.

70. Ralph Bienfang, *The Subtle Sense* (Norman: University of Oklahoma Press, 1946), p. 44.

4. VOICES, SOUNDS, AND HEAVENLY MUSIC

1. John Updike, *Self-Consciousness* (New York: Alfred A. Knopf, 1990), p. 233.

2. Robert Rivlin and Karen Gravelle, *Deciphering the Senses* (New York: Simon and Schuster, 1984), p. 78; Gina Kolata, "Studying Learning in the Womb," *Science*, 225 (20 July 1984), 302–303.

3. Desmond Morris, *The Naked Ape* (London: Corgi Books, 1968), pp. 94–96.

4. R. Murray Schafer, *The Tuning of the World* (New York: Alfred A. Knopf, 1977), pp. 226–227.

5. Quoted in Max Picard, *The World of Silence* (Chicago: Gateway/Henry Regnery, 1952), p. 39.

6. Thomas Nuttall, *A Journal of Travels into the Arkansa Territory, During the Year 1819* . . . (Philadelphia, 1821), in R. G. Thwaites, *Early Western Travels*, vol. 13, pp. 80–81, 93, 205; quoted in Howard Mumford Jones, *O Strange New World* (New York: Viking, 1964), p. 372.

7. Richard E. Byrd, *Alone* (Los Angeles: Tarcher, 1986), p. 119.

8. Schafer, *The Tuning of the World*, p. 23.

9. Ernest Shackleton, *South* (London: Heinemann, 1920), p. 40.

10. Gilbert C. Klingel, *Inagua* (London: Robert Hale); quoted in Margaret S. Anderson, *Splendour of Earth: An Anthology of Travel* (London: George Philip, 1963), p. 67.

11. Schafer, *The Tuning of the World*, pp. 35–36.

12. Alex Shoumatoff, "The Ituri Forest," *New Yorker*, February 6, 1984, p. 90.

13. Schafer, *The Tuning of the World*, p. 22.

14. Maxim Gorky, *Childhood*, quoted in Marco Valsecchi, trans. Arthur Coppotelli, *Landscape Painting of the Nineteenth Century* (Greenwich, Conn.: New York Graphic Society, 1971), p. 279.

15. Boris Pasternak, *Doctor Zhivago* (New York: Ballantine Books, 1958), p. 11; quoted in Schafer, *The Tuning of the World*, p. 32.

16. Leo Tolstoy, *Anna Karenina* (New York: Signet Classic, 1961), p. 165.

17. Ibid., p. 175.

18. Quoted in Ronald Blythe, *The Pleasures of Diaries* (New York: Pantheon, 1989), p. 101.

19. *The Epic of Gilgamesh*, ed. N. K. Sandars (Harmondsworth: Penguin, 1960), p. 105.

20. Juvenal, *Satires* III.236–259; Jerome Carcopino, *Daily Life in Ancient Rome: The People and the City at the Height of the Empire* (New Haven: Yale University Press, 1940), pp. 50, 180.

21. Quotations in Carl Bridenbaugh, *Cities in Revolt: Urban Life in America, 1743–1776* (New York: Alfred A. Knopf, 1955), p. 243.

22. Rosamond Bayne-Powell, *Eighteenth-Century London Life* (London: John Murray, 1937), p. 33.

23. John Betjeman, *Victorian and Edwardian London* (London: B. T. Batsford, 1969), pp. ix–xi.

24. Sheldon Cohen, "Sound Effects on Behavior," *Psychology Today*, October 1981, pp. 38–49.

25. Steen Eiler Rasmussen, *Experiencing Architecture* (Cambridge, Mass.: MIT Press, 1964), p. 225.

26. Michael Southworth, "The Sonic Environment of Cities," *Environment and Behavior*, 1, no. 1 (1969), 49–70; see also J. Douglas Porteous and Jane F. Mastin, "Soundscape," *Journal of Architectural Planning and Research*, 2 (1985), 169–186.

27. George S. Welsh, "The Perception of Our Urban Environment," in *Perceptions and Environment: Foundations of Urban Design*, ed. Robert E. Stipe (Chapel Hill: Institute of Government, University of North Carolina, 1966), p. 8.

28. Jacques Le Goff, *Time, Work, and Culture in the Middle Ages* (Chicago: University of Chicago Press, 1980), p. 46.

29. Max Picard, *The World of Silence* (Chicago: Gateway/Henry Regnery, 1952).

30. Colin Turnbull, "Liminality: A Synthesis of Subjective and Objective Experience," in *By Means of Performance: Intercultural Studies of Theatre and Ritual*, ed. Richard Schechner and Willa Appel (Cambridge: Cambridge University Press, 1990), pp. 54, 56, 58.

31. Colin Turnbull, "Legends of the BaMbuti," *Journal of the Royal Anthropological Institute*, 89 (1959), 45–60.

32. St. John Chrysostom, *Exposition of Psalm XLI*, translated in *Source Readings in Music History*, ed. Oliver Strunk (New York: W. W. Norton, 1950), pp. 69–70.

33. Victor Zuckerkandl, *Man the Musician* (Princeton: Princeton University Press, 1976), pp. 12–13.

34. Anton Chekhov, *Selected Letters*, ed. L. Hellman (New York: Farrar, Straus & Young, 1955), p. 142.

35. Bliss Wiant, *The Music of China* (Hong Kong: Chung Chi Publication, Chinese University of Hong Kong, n.d.), p. 7.

36. Li Ki [Chi], "Yueh Chi," book XVII, in *The Sacred Books of the East*, trans. James Legge (Oxford: Clarendon Press, 1884), vol. 28, p. 115.

37. Julius Portnoy, *The Philosopher and Music* (New York: Humanities Press, 1954), pp. 4–44.

38. Cicero, *De Re Publica* VI.8, trans. C. W. Keyes (Cambridge, Mass.: Harvard University Press, Loeb Library, 1928), pp. 271–273.

39. Plato, *Republic* 617a–b, trans. Benjamin Jowett; quoted in John Hollander, *The Untuning of the Sky: Ideas of Music in English Poetry, 1500–1700* (Princeton: Princeton University Press, 1961), p. 29.

40. Macrobius, *Commentary on the Dream of Scipio*, trans. W. H. Stahl (New York: Columbia University Press, 1952), p. 195; quoted in Hollander, *The Untuning of the Sky*, p. 30.

41. A. C. Schuldt, "The Voices of Time," *American Scholar*, Autumn 1976, pp. 554, 549–559.

42. Schafer, *The Tuning of the World*, p. 117.

43. Edward Rothstein, "Beethoven at Dusk," *New Republic*, March 21, 1988, p. 28.

44. Pablo Casals, *Joys and Sorrows: Reflections*, as told to Albert E. Kahn (New York: Simon & Schuster, 1970), p. 17.

45. George Steiner, "Orpheus with His Myths," in *Claude Lévi-Strauss: The Anthropologist as Hero*, ed. E. Nelson Hayes and Tanya Hayes (Cambridge, Mass.: MIT Press, 1970), p. 182.

46. George Steiner, Review of Brian McGuinness's *Wittgenstein: A Life*, in *London Review of Books*, June 23, 1988.

47. Peter Kivy, *Music Alone: Philosophical Reflections on the Purely Musical Experience* (Ithaca: Cornell University Press, 1990).

48. Gerald Brenan, *Thoughts in a Dry Season* (Cambridge: Cambridge University Press, 1978), p. 77.

49. Kivy, *Music Alone*, p. 88.

50. Lawrence Weschler, "Boy Wonder," *New Yorker*, November 18, 1986, pp. 88–89.

5. VISUAL DELIGHT AND SPLENDOR

1. Bernard G. Campbell, *Human Evolution: An Introduction to Man's Adaptations* (Chicago: Aldine, 1974).

2. E. H. Gombrich, *The Sense of Order: A Study in the Psychology of Decorative Art* (Ithaca: Cornell University Press, 1984).

3. Colin M. Turnbull, *The Forest People* (London: Chatto & Windus, 1961), p. 228.

4. Oliver Sacks, *The Man Who Mistook His Wife for a Hat* (New York: Harper & Row, 1987), p. 199.

5. Claude Lévi-Strauss, *Myth and Meaning* (New York: Schocken Books, 1979), p. 18.

6. Bernard Berenson, *Seeing and Knowing* (Greenwich, Conn.: New York Graphic Society, 1953), p. 23.

7. Raymond Firth, *We, the Tikopeia* (London: George Allen & Unwin, 1957), p. 29.

8. Oliver Sacks and Robert Wasserman, "The Case of the Colorblind Painter," *New York Review of Books*, November 19, 1987, pp. 25–34.

9. Brent Berlin and Paul Kay, *Basic Color Terms: Their Universality and Evolution* (Berkeley: University of California Press, 1969).

10. Quoted in Peter Brown, *Augustine of Hippo* (Berkeley: University of California Press, 1969), pp. 180, 329.

11. Simone de Beauvoir, *Force of Circumstance*, trans. Richard Howard (London: Weidenfeld & Nicolson, 1965), pp. 206–207.

12. Johan Huizinga, *The Waning of the Middle Ages* (Garden City, N.Y.: Doubleday Anchor, 1954), pp. 269–270.

13. Georges Duby, *The Age of the Cathedrals: Art and Society, 980–1420* (Chicago: University of Chicago Press, 1981), p. 148.

14. Richard Sennett, *The Fall of Public Man* (Cambridge: Cambridge University Press, 1976), p. 163.

15. Alasdair Clayre, ed., *Nature and Industrialization: An Anthology* (Oxford: Oxford University Press, 1977), pp. 128–129.

16. Asa Briggs, *Iron Bridge to Crystal Palace: Impact and Images of the Industrial Revolution* (London: Thames and Hudson, 1979); Stephen Daniels, "Loutherbourg's Chemical Theatre: *Coalbrookdale by Night*," in *Painting and the Politics of Culture*, ed. John Barrell (Oxford: Oxford University Press, 1992), pp. 195–230.

17. Mark Stevens, "Church's Church," *New Republic*, January 8 and 15, 1990, pp. 30–32.

18. George Talbot, *At Home: Domestic Life in the Post-Centennial Era, 1876–1920* (Madison: State Historical Society of Wisconsin, 1976), p. 15.

19. C. S. Lewis, "The Shoddy Lands," in *Of Other Worlds*, ed. Walter Hooper (New York: Harper, Brace & World, 1966), pp. 99–106.

20. Quoted in Georges Poulet, *Studies in Human Time* (Baltimore: Johns Hopkins Press, 1956), p. 249.

21. Quoted in Peter Quennell, *The Pursuit of Happiness* (Boston: Little, Brown, 1988), p. 80.

22. Robert Bernard Martin, *Gerard Manley Hopkins: A Very Private Life* (New York: Putnam's, 1991), p. 190.

23. John Updike, "Monet Isn't Everything," *New Republic*, March 19, 1990, p. 28.

24. Suzannah Lessard, "Kinds of Places," *New Yorker*, October 14, 1985, p. 55.

25. Viktor E. Frankl, *Man's Search for Meaning* (New York: Washington Square Press, 1963), pp. 62–63.

26. Quoted in Aristotle, *Ethica Eudemia*, 1216a.

27. Hans Jonas, *The Gnostic Religion* (Boston: Beacon Press, 1963), pp. 257–258.

28. C. S. Lewis, *The Discarded Image* (Cambridge: Cambridge University Press, 1964), p. 55.

29. *Iliad*, bk. 18, in *Greek Literature in Translation*, ed. G. Howe, G. A. Harrer, and P. H. Epps (New York: Harper & Row, 1948), pp. 28–29.

30. Gilbert Highet, *Poets in a Landscape* (New York: Alfred A. Knopf, 1957).

31. English translation of Tao Yuan-ming's poem in Robert Payne, ed., *The White Pony: An Anthology of Chinese Poetry* (New York: Mentor Books, 1960), p. 140.

32. Leo Marx, *The Machine in the Garden: Technology and the Pastoral Ideal in America* (New York: Oxford University Press, 1964).

33. James Dougherty, *The Fivesquare City: The City in the Religious Imagination* (Notre Dame: University of Notre Dame Press, 1980); John S. Dunne, *The City of the Gods: A Study in Myth and Mortality* (Notre Dame: University of Notre Dame Press, 1978); Helen Rosenau, *The Ideal City: Its Architectural Evolution* (New York: Harper & Row, 1972); Paul Wheatley, *The Pivot of the Four Quarters* (Chicago: Aldine, 1971).

34. Paul Lavedan, "Les Hittites et la cité circulaire," in *Histoire de l'urbanisme* (Paris: Henry Laurens, 1926), vol. 1, pp. 56–63; Guy Le Strange, *Baghdad during the Abbasid Caliphate from Contemporary and Persian Sources* (Oxford: Clarendon Press, 1924); A. F. Wright, "Symbolism and Function: Reflections on Ch'ang-an and Other Great Cities," *Journal of Asian Studies*, 24 (1965), 667–679.

35. Diana Eck, "The City as a Sacred Center," in *The City as a Sacred Center: Essays on Six Asian Contexts*, ed. Bardwell Smith and Holly Baker Reynolds (Leiden: E. J. Brill, 1987), pp. 5, 7.

36. Thomas Sharp, *Oxford Replanned* (London: Architectural Press, 1948), p. 32.

37. Rudolf Otto, *The Idea of the Holy* (London: Oxford University Press, 1958), pp. 12–24.

38. John K. Wright, "The Open Polar Sea," *Geographical Review*, 43 (1953), 338–365.

39. L. P. Kirwan, *A History of Polar Exploration* (Harmondsworth: Penguin Books, 1962); Chauncy C. Loomis, "The Arctic Sublime," in *Nature and the Victorian Imagination*, ed. U. C. Knoepflmacher and G. B. Tennson (Berkeley: University of California Press, 1977), pp. 95–112.

40. Fridtjof Nansen, *Farthest North: Being the Record of a Voyage of Exploration of the Ship "Fram" 1893–96* . . . (New York: Harper & Brothers, 1897), vol. 2, pp. 446–447.

41. Fridtjof Nansen, *The First Crossing of Greenland* (London: Longmans, 1892), p. 313.

42. Nansen, *Farthest North*, vol. 1, p. 1.

43. Ibid., vol. 2, p. 41.

44. Richard E. Byrd, *Discovery* (New York: Putnam's, 1935), p. 167; idem, *Alone* (1938; reprint, Los Angeles: Tarcher, n.d.), p. 178.
45. Byrd, *Alone*, pp. 25, 73–74, 85.
46. Nansen, *Farthest North*, vol. 2, p. 446.
47. Byrd, *Alone*, p. 179.

6. AUSTRALIAN ABORIGINES, THE CHINESE, AND MEDIEVAL EUROPEANS

1. Bernard Smith, *European Vision and the South Pacific, 1768–1850* (Oxford: Clarendon Press, 1960).
2. R. Brough Smyth, *The Aborigines of Victoria* (London: Trubner, 1878), vol. 1, p. 291.
3. A. P. Elkin, *The Australian Aborigines* (Garden City, N.Y.: Doubleday Anchor, 1964), p. 236.
4. Ronald M. Berndt and Catherine H. Berndt, *The World of the First Australians* (Chicago: University of Chicago Press, 1964), p. 95.
5. W. E. H. Stanner, *White Man Got No Dreaming: Essays, 1938–1973* (Canberra: Australia National University Press, 1979), pp. 38–39.
6. Elkin, *Australian Aborigines*, pp. 255, 271.
7. Ibid., p. 243.
8. Peter Sutton, ed., *Dreamings: The Art of Aboriginal Australia* (New York: George Braziller, 1989), pp. 96, 201.
9. Berndt and Berndt, *World of the First Australians*, p. 355.
10. Elkin, *Australian Aborigines*, p. 245.
11. Berndt and Berndt, *World of the First Australians*, pp. 314, 319.
12. T. G. H. Strehlow, *Aranda Traditions* (Carlton, Victoria: Melbourne University Press, 1947), pp. 26–30.
13. Bruce Chatwin, *The Songlines* (Harmondsworth: Penguin Books, 1988).
14. Elkin, *Australian Aborigines*, pp. 156–157.
15. David Hawkes, ed. and trans., *Ch'u Tz'u: The Songs of the South* (Boston: Beacon Paperback, 1962), pp. 119–120.
16. B. Karlgren, "Some Fecundity Symbols in Ancient China," *Bulletin of the Museum of Far Eastern Antiquities* (Stockholm), no. 2 (1930), 1–21; Maurice Freedman, "Geomancy and Ancestor Worship," in *Chinese Lineage and Society* (London: Athlone Press, 1966), pp. 124–127.
17. Michael Sullivan, *The Birth of Landscape Painting in China* (Berkeley: University of California Press, 1962).
18. Ibid., pp. 29–30.

19. T'ao Yuan-ming, "The Return," in *The White Pony*, trans. Robert Payne (New York: Mentor Books, 1960), p. 144.

20. Michael Sullivan, *Symbols of Eternity: The Art of Painting in China* (Stanford: Stanford University Press, 1979), pp. 80–81.

21. Quoted in Richard Edwards, *The World around the Chinese Artist: Aspects of Realism in Chinese Painting* (Ann Arbor: University of Michigan Press, 1989), p. 68.

22. Quoted in James C. Y. Watt, "The Literati Environment," in Chu-tsing Li and James C. Y. Watt, *The Chinese Scholar's Studio: Artistic Life in the Late Ming Period* (New York: Asia Society/Thames and Hudson, 1987), p. 5.

23. Joanna F. Handlin Smith, "Gardens in Ch'i Piao-chia's Social World: Wealth and Values in Late-Ming Kiangnan," *Journal of Asian Studies*, 51, no. 1 (1992), 55–81.

24. Maggie Keswick, "Foreword," in Ji Cheng, *The Craft of Gardens* (New Haven: Yale University Press, 1988), p. 23.

25. Sullivan, *Symbols of Eternity*, p. 26.

26. Watt, "The Literati Environment," p. 17.

27. Chu-tsing Li, "The Artistic Theories of the Literati," in Li and Watt, *The Chinese Scholar's Studio*, p. 20.

28. Arthur de Carle Sowerby, *Nature in Chinese Art* (New York: John Day, 1940), pp. 153–160.

29. Sullivan, *Symbols of Eternity*, pp. 57, 69–70.

30. Edwards, *World around the Chinese Artist*, p. 86.

31. Ibid., p. 91.

32. Ibid., pp. 110, 112.

33. Ibid., pp. 24, 29.

34. Chu-tsing Li, "Artistic Theories of the Literati," p. 21.

35. Edwards, *World around the Chinese Artist*, pp. 146–147.

36. Derek Pearsall and Elizabeth Salter, *Landscapes and Seasons of the Medieval World* (London: Paul Elek, 1973).

37. C. S. Lewis, *The Discarded Image: An Introduction to Medieval and Renaissance Literature* (Cambridge: Cambridge University Press, 1964), p. 101.

38. Georges Duby, *The Age of the Cathedrals: Art and Society, 980–1420*, trans. Eleanor Levieux and Barbara Thompson (Chicago: University of Chicago Press, 1981), p. 152.

39. Quoted in Edward A. Armstrong, *Saint Francis: Nature Mystic* (Berkeley: University of California Press, 1976), p. 9.

40. Quoted in Duby, *The Age of the Cathedrals*, p. 210.

41. Lewis, *The Discarded Image*, pp. 98–100.

42. Umberto Eco, *Art and Beauty in the Middle Ages*, trans. Hugh Bredin (New Haven: Yale University Press, 1986), p. 32.

43. Quoted in Lewis, *The Discarded Image*, pp. 111–112.

44. Jean Gimpel, *The Cathedral Builders* (New York: Harper Colophon Books, 1984), pp. 41–42.

45. Quoted in Duby, *The Age of the Cathedrals*, pp. 80–81.

46. Johan Huizinga, *The Waning of the Middle Ages* (Garden City, N.Y.: Doubleday Anchor, 1954), pp. 270–273.

47. Eco, *Art and Beauty*, pp. 44–46.

48. Otto von Simson, *The Gothic Cathedral: Origins of Gothic Architecture and the Medieval Concept of Order* (New York: Pantheon Books, 1962), p. 52.

49. Duby, *The Age of the Cathedrals*, p. 102.

50. Patrick Nuttgens, *The Landscape of Ideas* (London: Faber and Faber, 1972), p. 60.

51. J. R. Johnson, *The Radiance of Chartres*, Columbia University Studies in Art and Archaeology no. 4 (New York: Phaidon, 1964), pp. 56–57.

52. Erwin Panofsky, *Abbot Suger on the Abbey Church of St.-Denis and Its Art Treasures* (Princeton: Princeton University Press, 1946), p. 14.

53. Quoted in ibid., p. 63.

54. Ibid., pp. 63, 65.

55. Quoted in Duby, *The Age of the Cathedrals*, p. 89.

56. Panofsky, *Abbot Suger on St.-Denis*, p. 19.

57. Simson, *The Gothic Cathedral*, pp. 3–4.

58. Quoted in Panofsky, *Abbot Suger on St.-Denis*, p. 22.

7. AMERICAN PLACE AND SCENE

1. Helen Hooven Santmyer, *Ohio Town* (Columbus: Ohio State University Press, 1962), pp. 308–309.

2. Ibid., p. 50.

3. Paul Horgan, *Whitewater* (New York: Paperback Library, 1971), p. 163.

4. John R. Stilgoe, "Fair Fields and Blasted Rock: American Land Classification Systems and Landscape Aesthetics," *American Studies*, 22 (Spring 1981), 21–33; Dwight quoted in Allen Carlson, "On Appreciating Agricultural Landscapes," *Journal of Aesthetics and Art Criticism*, 43, no. 3 (Spring 1985), 301–311.

5. *The Writings of Colonel William Byrd of Westover in Virginia Esqr'*, ed. John Spencer Bassett (New York: Doubleday, 1901), pp. 135, 146, 163, 172, 186.

6. Thomas Jefferson, *Notes on the State of Virginia*, ed. William Peden (Chapel Hill: University of North Carolina, 1955), p. 19.

7. Roderick Nash, *Wilderness and the American Mind* (New Haven: Yale University Press, 1967), p. 60.

8. Mark Stevens, "Church's Church," *New Republic*, January 8 and 15, 1990, pp. 30–32; Yi-Fu Tuan, "Paradoxical Images of the American West," in Ellen M. Murgy and Jeane M. Knapp, *Kaleidoscope of History*, American Geographical Society Collection Special Publication no. 1 (Milwaukee: University of Wisconsin, 1990), pp. 104–106.

9. Howard Mumford Jones, *O Strange New World* (New York: Viking, 1964), p. 270.

10. Hildegard Binder Johnson, *The Orderly Landscape: Landscape Tastes and the United States Survey*, James Ford Bell Lectures no. 15 (Minneapolis: University of Minnesota, 1977); Robert David Sack, *Human Territoriality: Its Theory and History* (Cambridge: Cambridge University Press, 1986), pp. 144–163.

11. J. B. Jackson, *Discovering the Vernacular Landscape* (New Haven: Yale University Press, 1984), p. 67.

12. St. John de Crèvecoeur, *Letters from an American Farmer* (London: J. M. Dent; New York: E. P. Dutton, 1912), p. 11.

13. *Landscapes: Selected Writing of J. B. Jackson*, ed. Ervin H. Zube (Amherst: University of Massachusetts Press, 1970), pp. 47–48.

14. Perry Miller, *Errand into Wilderness* (New York: Harper Torchbooks, 1964), pp. 210, 211–212.

15. Quotations in Michael H. Cowan, *City of the West: Emerson, America, and Urban Metaphor* (New Haven: Yale University Press, 1967), pp. 44–47.

16. David Lowenthal, "The Past in the American Landscape," in *Geographies of the Mind*, ed. David Lowenthal and Martyn J. Bowden (New York: Oxford University Press, 1976), p. 91.

17. Quotations in ibid., pp. 91, 93, 94–95.

18. Ned Rorem, *The Final Diary* (New York: Holt, Rinehart and Winston, 1974), p. 146.

19. Quotations in Susan Edmiston and Linda D. Cirino, eds., *Literary New York: A History and Guide* (Boston: Houghton Mifflin, 1976), pp. 251–252.

20. *New Yorker*, July 27, 1981, p. 25.

21. David Lowenthal, "The American Scene," *Geographical Review*, 58 (1968), 69. The two observers are, respectively, John A. Kouwenhoven and David Lowenthal.

22. Ibid., p. 77.

23. Edmund Wilson, *I Thought of Daisy*; quoted by John Updike, *Hugging the Shore* (New York: Alfred A. Knopf, 1983), p. 204.

24. John Cheever, *Bullet Park* (New York: Alfred A. Knopf, 1969), pp. 3–4.

25. Tom McKnight, "Irrigation Technology: A Photo-Essay," *Focus*, 40 (Summer 1990), 1–6.

26. Jackson, *Landscapes*, p. 57.

27. J. B. Jackson, "The Abstract World of the Hot-Rodder," *Landscape*, 7 (Winter 1957–58), 22–27.

28. *The Prose of Philip Freneau*, ed. Philip M. Marsh (New Brunswick, N.J.: Scarecrow Press, 1955), p. 228.

29. Nash, *Wilderness and the American Mind*, pp. 108, 113.

30. Hans Huth, *Nature and the American: Three Centuries of Changing Attitudes* (Lincoln: University of Nebraska Press, 1972), p. 137.

31. Karal Ann Marling, *The Colossus of Roads: Myth and Symbol along the American Highway* (Minneapolis: University of Minnesota Press, 1984), pp. 2–3.

32. Jackson, *Landscapes*, pp. 64–65.

33. Ibid., p. 66.

34. Robert Venturi, Denise Scott Brown, and Steven Izenour, *Learning from Las Vegas* (Cambridge, Mass.: MIT Press, 1972), p. 31.

8. SYNESTHESIA, METAPHOR, AND SYMBOLIC SPACE

1. John Cowper Powys, *The Art of Growing Old* (London: Jonathan Cape, 1944), p. 100.

2. T. F. Karwoski and H. S. Odbert, "Color-Music," *Psychological Monographs*, 50, no. 2 (1938), 3.

3. Lawrence E. Marks, "Synesthesia," *Psychology Today*, 9, no. 1 (1975), 48–52; *The Unity of the Senses: Interrelations among the Modalities* (New York: Academic Press, 1978).

4. Philip Wheelwright, *Metaphor and Reality* (Bloomington: Indiana University Press, 1962), p. 76.

5. Vladimir Nabokov, *Speak, Memory* (New York: Putnam's, 1966), pp. 34–35.

6. A. R. Luria, *The Mind of a Mnemonist* (New York: Basic Books, 1968), pp. 24, 38.

7. Charles E. Osgood, "The Cross-Cultural Generality of Visual-Verbal Synesthetic Tendencies," *Behavioral Science*, 5, no. 2 (1960), 146–169.

8. Ibid., p. 168; see also Charles E. Osgood, William H. May, and Murray S. Miron, *Cross-Cultural Universals of Affective Meaning* (Urbana: University of Illinois Press, 1975), pp. 397–399.

9. Howard Gardner, *Art, Mind, and Brain* (New York: Basic Books, 1982), p. 99.

10. James Fernandez, "The Mission of Metaphor in Expressive Culture," *Current Anthropology*, 15, no. 2 (1974), 122; see also Cecil H. Brown and Stanley R. Witkowski, "Figurative Language in a Universalist Perspective," *American Ethnologist*, 8, no. 3 (1981), 596–615.

11. Jane H. Stolper, "Color Induced Physiological Response," *Man-Environment Systems*, 7 (1977), 101–108.

12. Yi-Fu Tuan, "Sign and Metaphor," *Annals of the Association of American Geographers*, 68, no. 3 (1978), 363–372.

13. Robert David Sack, *Conceptions of Space in Social Thought: A Geographic Perspective* (London: Macmillan, 1980).

14. Karl A. Nowotny, *Beiträge zur Geschichte des Weltbildes* (Vienna: Ferdinand Berger, 1970).

15. John G. Neihardt, *Black Elk Speaks: Being the Life Story of a Holy Man of the Oglala Sioux* (Lincoln: University of Nebraska Press, 1961), p. 279.

16. Ibid., pp. 198–199, 277–279.

17. Alfred Forke, *The World-Conception of the Chinese* (London: Arthur Probsthain, 1925); Marcel Granet, *La Pensée Chinoise* (Paris: Albin Michel, 1934), especially the section "Le Microcosme," pp. 361–388; see also John B. Henderson, *The Development and Decline of Chinese Cosmology* (New York: Columbia University Press, 1984).

18. Joseph Needham, *Science and Civilisation in China* (Cambridge: Cambridge University Press, 1956), vol. 2, p. 261.

19. Translation in E. R. Hughes, *Chinese Philosophy in Classical Times* (London: J. M. Dent, 1942), p. 294.

20. Nelson I. Wu, *Chinese and Indian Architecture* (New York: George Braziller, 1963), pp. 29–45; Arthur F. Wright, "Symbolism and Function: Reflections on Changan and Other Great Cities," *Journal of Asian Studies*, 24 (1965), 670–671.

21. Russel Ward, *The Australian Legend* (Melbourne: Oxford University Press, 1966).

22. "The middle is seen as more typical than the periphery, the small community more typical than the metropolis. Taken to an extreme—finding the middle of the middle of the middle—the very essence of American life ought to be found in a small town located in the middle of the state of Iowa"; John C. Hudson, Review of Thomas J. Morain, *Prairie Grass Roots: An Iowa Small Town in the Early Twentieth Century*, *Indiana Magazine of History*, December 1989, p. 364; see also James R. Shortridge, "The Vernacular Middle West," *Annals of the Association of American Geographers*, 75, no. 1 (1985), 48–57.

23. J. B. Jackson, *American Space: The Centennial Years, 1865–1876* (New York: Norton, 1970), p. 58.

24. Howard Mumford Jones, *The Age of Energy: Varieties of American Experience, 1865–1915* (New York: Viking Press, 1971), pp. 71–72.

25. Leslie Fiedler, *The Return of the Vanishing American* (New York: Stein & Day, 1968), pp. 16–22.

9. RITUAL AND THE AESTHETIC-MORAL STATE

1. Jacques Soustelle, *Daily Life of the Aztecs on the Eve of the Spanish Conquest* (Stanford: Stanford University Press, 1970), pp. 95–102; Bernard R. Ortiz de Montellano, "Aztec Cannibalism: An Ecological Necessity?" *Science*, May 12, 1978, pp. 611–617.

2. Wolfram Eberhard, *A History of China*, 2d ed. (Berkeley: University of California Press, 1960), p. 23.

3. Marcel Granet, *Chinese Civilization* (New York: Meridian Books, 1958), pp. 191, 208.

4. Louis A. Hieb, "Meaning and Mismeaning toward an Understanding of the Ritual Clown," in *New Perspectives on the Pueblos*, ed. Alfonso Ortiz (Albuquerque: University of New Mexico Press, 1972), pp. 171–187.

5. Mikhail Bakhtin, *Rabelais and His World* (Cambridge, Mass.: MIT Press, 1968), pp. 19–21.

6. Benjamin I. Schwartz, "Transcendence in Ancient China," *Daedalus*, 104 (Spring 1975), 58–59.

7. David I. Kertzer, *Ritual, Politics, and Power* (New Haven: Yale University Press, 1988).

8. Nelson I. Wu, *Chinese and Indian Architecture: The City of Man, the*

Mountain of God, and the Realm of the Immortals (New York: Braziller, 1963), pp. 37–38.

9. Arthur F. Wright, *The Sui Dynasty: The Unification of China, A.D. 581–617* (New York: Alfred A. Knopf, 1978), pp. 87–88.

10. Arthur F. Wright, "Symbolism and Function: Reflections on Ch'ang-an and Other Great Cities," *Journal of Asian Studies*, 23 (1965), 667–679.

11. W. E. Soothill, *The Hall of Light: A Study of Early Chinese Kingship* (London: Lutterworth Press, 1951), pp. 25–29.

12. Howard J. Wechsler, *Offerings of Jade and Silk: Ritual and Symbol in the Legitimation of the T'ang Dynasty* (New Haven: Yale University Press, 1985), p. 210.

13. *Hsun Tzu: Basic Writings*, trans. Burton Watson (New York: Columbia University Press, 1963), pp. 109–110.

14. Ibid., p. 89.

15. Wechsler, *Offerings of Jade and Silk*, p. 29.

16. David McMullen, "Bureaucrats and Cosmology: The Ritual Code of T'ang China," in *Rituals of Royalty: Power and Ceremonial in Traditional Societies*, ed. David Cannadine and Simon Price (Cambridge: Cambridge University Press, 1987), p. 219.

17. Colin Morris, *The Discovery of the Individual, 1050–1200* (New York: Harper Torchbook, 1973), p. 25.

18. Janet L. Nelson, "The Lord's Anointed and the People's Choice: Carolingian Royal Ritual," in Cannadine and Price, *Rituals of Royalty*, p. 143.

19. Edward Shils and Michael Young, "The Meaning of Coronation," *Sociological Review*, 1 (1953), 80.

20. Lauro Martines, *Power and Imagination: City-States in Renaissance Italy* (New York: Vintage Books, 1980), p. 232.

21. Edward Muir, *Civic Ritual in Renaissance Venice* (Princeton: Princeton University Press, 1981), pp. 15–16.

22. Denis Cosgrove, "Venice, the Veneto and Sixteenth Century Landscape," in *Social Formation and Symbolic Landscape* (Beckenham, Kent: Croom Helm, 1984), pp. 102–141.

23. Muir, *Civic Ritual in Renaissance Venice*, pp. 112, 209, 219.

24. Frederic C. Lane, *Venice: A Maritime Republic* (Baltimore: Johns Hopkins University Press, 1973), p. 271.

25. Ibid.

26. Muir, *Civic Ritual in Renaissance Venice*, p. 50.

27. Julia S. Berrall, *The Garden: An Illustrated History* (New York: Viking Press, 1966), p. 200.

28. Olivier Bernier, *Louis XIV: A Royal Life* (New York: Doubleday, 1987), pp. 212, 227.

29. Ibid., p. 97.

30. Ibid., pp. 101–102, 220.

31. W. H. Lewis, *The Splendid Century: Life in the France of Louis XIV* (New York: Morrow Quill Paperback, 1978), pp. 47, 202.

32. George Santayana, *The Sense of Beauty: Being the Outlines of Aesthetic Theory* (New York: Charles Scribner, 1896), pp. 84–85.

33. Daniel J. Boorstin, *The Americans: The National Experience* (New York: Vintage Books, 1965), p. 219.

34. Jacob Burckhardt, *The Civilization of the Renaissance in Italy* (1860; reprint, London: Penguin Books, 1990), pp. 19–97.

35. James Thomas Flexner, *George Washington and the New Nation, 1783–1793* (Boston: Little, Brown, 1970), pp. 182, 194.

36. John W. Reps, *Town Planning in Frontier America* (Princeton: Princeton University Press, 1969), pp. 304–343.

37. Hans Paul Caemmerer, *Washington: The National Capital City* (Washington, D.C.: U.S. Government Printing Office, 1932), p. 29.

38. Marshall Berman, *All That Is Solid Melts into Air: The Experience of Modernity* (New York: Penguin Books, 1988), p. 178.

39. Charles Hurd, *The White House: A Biography* (New York: Harper & Brothers, 1940), p. 125.

40. John Burchard and Albert Bush-Brown, *The Architecture of America: A Social and Cultural History* (Boston: Little, Brown, 1966), p. 21.

41. David Lowenthal, "The American Way of History," *Columbia University Forum*, 9, no. 3 (1966), 32.

42. St. John de Crèvecoeur, *Letters from an American Farmer* (London: T. Davies, 1782), pp. 46–48.

43. Quoted in Flexner, *George Washington and the New Nation*, p. 71.

44. W. Lloyd Warner, "An American Sacred Ceremony," in Russel E. Richey and Donald G. Jones, *American Civil Religion* (New York: Harper & Row, 1974), pp. 91–99.

45. Ibid., p. 99.

46. Boorstin, *The Americans*, p. 245; Robert David Sack, "The American Territorial System," in *Human Territoriality: Its Theory and History* (Cambridge: Cambridge University Press, 1986), pp. 127–168.

47. Burchard and Bush-Brown, *The Architecture of America*, p. 311.

48. Flexner, *George Washington and the New Nation*, p. 159.

49. James Madison, *The Federalist* no. 51 (February 19, 1788), reprinted in Robert C. Baron, ed., *Soul of America: Documenting Our Past, 1492–1974* (Golden, Colo.: Fulcrum, 1989), pp. 111–112; Arthur O. Lovejoy, *Reflections on Human Nature* (Baltimore: Johns Hopkins Press, 1961), pp. 37–65.

50. Quoted in Baron, *Soul of America*, p. 343.

10. GOOD AND BEAUTIFUL

1. Clyde Kluckhohn, "Expressive Activities," in Evon Vogt and Ethel Albert, *People of Rimrock: A Study of Values in Five Cultures* (Cambridge, Mass.: Harvard University Press, 1966), p. 283.

2. Roland H. Bainton, *Behold the Christ: A Portrayal of Christ in Words and Pictures* (New York: Harper & Row, 1974).

3. Camille Paglia, *Sexual Personae: Art and Decadence from Nefertiti to Emily Dickinson* (New York: Vintage Books, 1991), p. 594; Richard Chase, *Herman Melville: A Critical Study* (New York: Macmillan, 1949), p. 266.

4. Lewis Mumford, *The City in History* (New York: Harcourt, Brace & World, 1961), p. 137.

5. Walter L. Creese, *The Crowning of American Land: Eight Great Spaces and Their Buildings* (Princeton: Princeton University Press, 1985).

6. George H. Williams, *Paradise and Wilderness in Christian Thought* (New York: Harper & Row, 1962).

7. *Life and Works of Saint Bernard, Abbot of Clairvaux*, ed. J. Mabillon, trans. with additional notes S. E. Eales (London: J. Hodges, 1889), vol. 2, p. 464.

8. Interview with Malcolm Cowley in George Plimpton, ed., *Writers at Work: The Paris Interviews* (New York: Viking, 1986), pp. 19–20.

9. Wright Morris, *The Home Place* (New York: Scribner's, 1948), p. 143.

10. Brian McGuinness, *Wittgenstein, A Life: Young Ludwig, 1889–1921* (Berkeley: University of California Press, 1988), p. 252.

11. Alain [Emile Chartier], *On Happiness* (New York: Ungar, 1973), p. 243.

12. Quoted by Jonathan Schell, who himself writes: "The timeless appeal of the greatest works of art, in fact, testifies to our common humanity as few other things do, and is one of the strongest grounds we have for supposing that a political community that would embrace the whole

earth and all generations is also possible"; "The Fate of the Earth," pt. 2, *New Yorker*, February 8, 1982, p. 95.

13. Robert Grudin, *Time and the Art of Living* (New York: Ticknor & Fields, 1988), pp. 168–169.

14. Freya Stark, *The Journey's End* (London: John Murray, 1963), p. 220.

15. Calvin Tomkins, "Profile of John M. Brealey," *New Yorker*, March 16, 1987, pp. 50–51.

16. Iris Murdoch, *The Sovereignty of Good* (New York: Schocken Books, 1971), p. 88.

17. Peter Winch, *Simone Weil: "The Just Balance"* (Cambridge: Cambridge University Press, 1989), p. 173.

18. "That the good and true are one and the same is part of our heavy legacy from Plato; in India, too, the ideas merge so completely that one Sanskrit word (*sat*, related to our "is") means not only what is true, and what is good, but what is real"; Wendy Doniger O'Flaherty, "The Boundary between Myth and Reality," in *Intelligence and Imagination*, *Daedalus*, Spring 1980, pp. 104–105.

19. Iris Murdoch, *The Bell* (Harmondsworth: Penguin, 1962), pp. 191–192.

20. Quoted in Stephen Gill, *William Wordsworth: A Life* (Oxford: Clarendon Press, 1989), pp. 181–182.

21. Joyce Cary, *Art and Reality: Ways of the Creative Process* (Garden City, N.Y.: Doubleday Anchor, 1961), pp. 127, 161. Some authors insist, unconvincingly, that their works are wholly without moral implication, such as Vladimir Nabokov, *Strong Opinions* (New York: Vintage International, 1990); see also Tobin Siebers, *Morals and Stories* (New York: Columbia University Press, 1992).

22. Mark Stevens, "Retouching Rembrandt," *New Republic*, August 22, 1988, p. 29.

23. Mark Stevens, "The Unblinking Eye," *New Republic*, November 9, 1987, p. 33.

24. George Steiner, *Martin Heidegger* (Chicago: University of Chicago Press, 1987), p. 43.

25. Quotations in Johan Huizinga, *The Waning of the Middle Ages* (Garden City, N.Y.: Doubleday Anchor, 1954), p. 268.

26. McGuinness, *Wittgenstein*, p. 112.

27. Gerald Brenan, *Thoughts in a Dry Season* (Cambridge: Cambridge University Press, 1978), p. 77.

28. William Styron, *Darkness Visible: A Memoir of Madness* (New York: Random House, 1990), pp. 66–67.

29. Cary, *Art and Reality*, p. 161.

EPILOGUE

1. Kevin Duffy, *Children of the Forest* (New York: Dodd Mead & Co., 1984), pp. 161–166; on Roman and medieval orgies, see Reay Tannahill, *Food in History* (New York: Stein & Day, 1974).

2. Iskandar Carey, *Orang Asli: The Aboriginal Tribes of Peninsular Malaysia* (Kuala Lumpur: Oxford University Press, 1976), p. 99.

3. Joseph Levenson and Franz Schurmann, *China: An Interpretive History* (Berkeley: University of California Press, 1971), pp. 114–115.

4. Kenneth K. S. Chen, *Buddhism* (Woodbury, N.Y.: Barron's Educational Series, 1968), p. 149.

5. Lynn White, Jr., *Machina ex Deo* (Cambridge, Mass.: MIT Press, 1968), p. 63.

6. Pierre du Colombier, *Les Chantiers des cathédrales* (Paris: J. Picard, 1953), p. 18; quoted in Adolf Katzenellenbogen, *The Sculptural Programs of Chartres Cathedral* (New York: Norton, 1964), p. vii.

7. John Updike, "Our National Monument," in *Odd Jobs* (New York: Alfred A. Knopf, 1991), pp. 77–84.

8. E. E. LeMasters, *Blue-Collar Aristocrats: Life-Styles at a Working-Class Tavern* (Madison: University of Wisconsin Press, 1976), pp. 23–24.

9. Mencius, *The Four Books*, trans. James Legge (New York: Paragon, 1966), bk. 3, pt. 2, pp. 674–675; Raymond Williams, *The Country and the City* (New York: Oxford University Press, 1973), pp. 66, 82, 109.

10. Arthur F. Wright, *The Sui Dynasty* (New York: Alfred A. Knopf, 1978), pp. 49–50.

11. Gilette Ziegler, *The Court of Versailles in the Reign of Louis XIV* (London: George Allen & Unwin, 1966), p. 30.

12. Stuart Hampshire, "The Tory Anarchist," Review of Michael Shelden, *Orwell: The Authorized Biography* (HarperCollins, 1991), in *New York Review of Books*, January 30, 1992, p. 12.

13. Yi-Fu Tuan, "Geopiety: A Theme in Man's Attachment to Nature and to Place," in David Lowenthal and Martyn J. Bowden, *Geographies of the Mind* (New York: Oxford University Press, 1976), pp. 15–17; idem, *Morality and Imagination: Paradoxes of Progress* (Madison: University of Wisconsin Press, 1989), pp. 38–49.

14. E. M. W. Tillyard, *The Elizabethan World Picture* (London: Chatto & Windus, 1960), p. 24–25.

15. John P. Sisk, "The Tyranny of Harmony," *American Scholar*, Spring 1977, pp. 193–205; Barbara Miller Lane, *Architecture and Politics in Germany, 1918–1945* (Cambridge, Mass.: Harvard University Press, 1985); David I. Kertzer, *Ritual, Politics, and Power* (New Haven: Yale University Press, 1988), pp. 163–167.

A C K N O W L E D G M E N T S

Once again I would like to thank the Vilas Trust of the University of Wisconsin for supporting my work so generously during the last eight years. As for indebtedness to individuals, the longer one is in the writing business the greater and more numerous, of course, are the debts incurred. But how satisfying it is to be able to draw on other people's knowledge and insights—to be an intellectual debtor! Naming names does become more of a problem—more arbitrary—as time passes. Nevertheless, I would like to mention five scholars—Merle Curti, Nicholas Entrikin, David Lowenthal, Kenneth Olwig, and Robert Sack—from whom I have received much needed encouragement of both the direct and indirect kind, and four editors—Tom Engelhardt, Jack Kirschbaum, and (for this book) Howard Boyer and Ann Hawthorne. Editors, I have always known, can significantly improve a manuscript, but I never quite realized by how much until Howard and Ann wielded their blue pencils and attached yellow tabs on my work.

A
Aborigines, Australian, 122–27
  art and, 122–25
  dreaming and, 125–27
  ethos and aesthetics, 123–25
  symbolic space of, 177
Abstinence, 52–53
Abstract, aesthetics of the, 13–16
Achilles, 11
Adams, Ansel, 218
Adams, John, 201

Aerial views, 155
Aestheticism:
  Aboriginal, 122–25
  American, 143–61
  Chinese, 127–35
  consciousness, emotion, and, 7–9
  criticism of, 19
  development of the impulse of,
    20–31
  extreme, 19
  home and, 113–14

morality and, *see* Good and beautiful
practical life and, 100–101
see also Ritual and aesthetic-moral state
Aggressive commercialism, 157–61
Agriculture, 17–18, 110–11, 113–14, 145
  moralism and, 148–49
Album of Landscapes, 134
Alto Rhapsody, 226
America, 143–61
  aesthetic-moral state in, 199–209
  agricultural land, 145, 148–49
  amusing, gigantic, and bizarre, 156–58
  cities, 150–53
  Classical values, 147–48
  dignity of the state, 242–43
  home and hometown, 144–45
  landscapes, 145–46, 155–56, 242
  moralism, 148–51, 242
  movement, 154–56
  newness and cleanness, 151–53
  order on the land, 148
  process, 153–54
  rituals of state, 242
  Romanticism, 146–47
  the strip, 158–61
  symbolic space in, 178–81
  wilderness, 145–47, 149, 218
Anarchy, 185–86, 236
Anaxagoras, 109
Anderson, E. N., 54
Anger, 11
Animal cries, 76–77, 84
Animal epithets, 171

Animal instinct, 20–21
Animals, 229–30
  as food, 229
Ankor Thom, 112
Anna Karenina (Tolstoy), 38
Antarctica, 75, 76, 117
Anthropomorphics, 13
Aquinas, Saint Thomas, 139
Architecture:
  aromatic, 64–68
  city, 150–51
  for democracy, 201–203
  food and, 49–50
  human cost of, 230–33
Art and artists, 27–29, 102–108, 201, 218–20
  Aboriginal, 122–25
  Chinese, 127–35
  clarity of art, 221
  color and, 102–104
  controlling intelligence and, 200
  democracy as, 205
  gemlike fire versus twilight, 104–106
  gigantism and, 157
  intense and joyful seeing, 106–108
  preaching and, 223–24
  prestige and authority of art, 220–26
  principles of, 208–209
  reality and, 222–25
  transience and, 221–22
Artifacts, 219
Artisan-artist, 218–20, 234
Athletes, 36–37, 38
Auden, W. H., 221
Augustine, Saint, 66
Augustine of Hippo, 104
Aurora borealis, 116

Australia:
  Aborigines, *see* Aborigines, Australian
  symbolic space in, 177–78
Aztecs, 183–84

B
Babylonians, 64–65
Bach, Johann Sebastian, 95
Baltimore, 64
Banares, 111, 112
Bannister, Roger, 36–37
Barthes, Roland, 14
Beautiful country, 145
Beauty, 19, 139, 145
  children and, 21, 24, 25
  culture and, 101
  home and, 144–45
  moral, 240–43
  touch, reality, and, 45–46
  wilderness, 145–46
Beethoven, Ludwig van, 92, 93, 226, 230
Beginnings, 17
Bells, 83, 88
Berenson, Bernard, 43, 100
Berlin, Brent, 103
Bernard, Saint, 217–18
Bible, 60, 66, 74, 151, 191–92, 200, 201, 216, 227
Bicentennial, 242
Biological rhythms, 71–72
Bird with the Beautiful Song, 86
Black, 173, 175
Black Elk, 173, 174
Blast Furnace by Moonlight, 106
Blue Ridge Mountains, 146, 203
Bodin, Jean, 195, 196, 200
Body, the, 228
Boileau-Despréaux, Nicolas, 48

Bonding, 56–57
Book of Rites, 187–89
Boorstin, Daniel, 200, 206
Boston, 82–83
Bottom-up process, 206
Brahms, Johannes, 226
Breathing, 72
Brenan, Gerald, 93–94, 226
Bruner, J. S., 21
Buddhism, 66, 67, 128, 130, 135, 215, 228, 229, 231, 232, 241
Buenaventure, Saint, 139
Burchard, John, 207
Burckhardt, Jacob, 201
Bush-Brown, Albert, 207
Butler, Samuel, 46
Byrd, Richard E., 75, 115, 117, 118
Byrd, William, 146

C
Camus, Albert, 37
Cannibalism, 11
Canticle of the Creatures, 137
Carcasses, 224
Cardinal points, 173–81, 215
Carême, Antonin, 49
Carolingians, 192
Carter, Jimmy and Rosalyn, 242
Cary, Joyce, 223, 226
Casals, Pablo, 92
Cassian, John, 217
"Casual chaos," 154
Cathedrals, Gothic, 137–42
Cedarwood, 65
Celestial Jerusalem, 138–39
Cézanne, Paul, 107, 108
Chain of being, 237
Chalcidius, 110

Chandrasekhar, S., 16
Ch'ang-an, 188
Change, 190–91
  without loss of essential form,
    208–209
Chang Tai, 54–55
Chan-tse-tuan, 133
Chaos, 236
Chardin, Pierre Teilhard de, 24
Charles the Bald, 192
Charles X, 192
Chartier, Emile, 220
Chartres, 231
Chatwin, Bruce, 126
Chaucer, Geoffrey, 136
Chekhov, Anton, 88
Children, 20–31
  capacity for wonder, 23
  color and, 103
  education of, 228
  growth in competence (music and
    visual art), 25–29
  growth in sensibility, 29–31
  metaphor and, 170
  remembrance of things past, 23–
    25
  smell and, 56–57, 59
  the strip and, 160
  synesthesia and, 170
  timelessness and sensory delight,
    22–23
  touch and, 41, 42
China, 235
  aesthetic-moral state in, 186–
    91, 215–16, 238, 239, 242
  cities, 111, 112, 187–88
  eating and manners, 46, 51–55,
    230
  family in, 237

farms, 111
gardens, 67, 131–32, 215, 232
music, 88–89
nature and landscape, 127–35
odor and, 60, 61, 65–66
symbolic space, 174–77, 179,
  180, 187–88
Ch'ing dynasty, 134
Chou dynasty, 184
Christ, 191, 192, 194, 198, 216,
  224
Christianity, 66, 109–10, 136–
  42, 198, 217–18, 224, 232
Christian kingship, 191–93
Chrysostom, St. John, 86
Church, Frederick, 106, 147
Cicero, 89
Cincirli, 111
Cistercians, 217–18
Cities, 18, 44, 131, 135
  American, 150–53
  architecture of, 150–51
  beautiful, 151–53
  fragrances in, 63–66
  idealization of, 150
  shapes of, 111–13
  sounds of, 79–84
Civilization, 129, 149, 187
Clairvaux, Abbey of, 218
Classical values, 147–48
Claudius, Matthias, 74
Cleanness, 152–53
Clowns, 185–86
Coalbrookdale by Night, 106
Coe, Richard, 23
Cognition and pure music, 93–95
Color, 102–104, 224
  cardinal points and, 173–76,
    180

cathedrals and, 139–42
metaphor and, 171
-sound synesthesia, 168
sparkle and, 23–25
Colorblindness, 102–103
Communal celebration, 85–87
Composition versus pattern, 97–98
Confucius, Confucianism, 51–52, 88–89, 111, 186, 215, 216, 237
Congress, 201, 202, 203
Conjunction of incommensurables, 222
Consciousness, 7–9
Consensus, 205–209
Constable, John, 43
Constitution, American, 195, 206–209, 243
Contemplation, 114–15
Copenhagen, 81–82
Cos, 217
Cosmos (cosmic):
  anarchy, 236
  cities, 111–12
  clock, 176–77
  harmony, 187
  insecurity, 183–84
  microcosm and, 108–13
  moral-aesthetic, 215–16
  space, 187–90
Countryside, fragrances in, 61–63
Courtesy, 198–99
Cowley, Malcolm, 219
Crabbe, George, 232
Creative enterprise, 233–34
Crèvecoeur, St. John de, 149, 204, 205
Csikszentmihalyi, Mihaly, 37

Culture:
  aesthetic in life and, 5–19
  color and, 103–104
  eating, taste, and, 46–55
  history of Western, 11–12, 16
  as material transformation, 16–18
  meanings of, 5–7
  moral beauty and, 240–43
  nature and, 7–8, 227–29
  as perception, 7
  as performance, 7
  as physical process, 6–7
  sight and, 98–102
  as speech, 7

D
D'Ailly, Pierre, 226
Dance, 38–39
Dante, 66, 138
David, 191
Da Vinci, Leonardo, 216
Death, 73
  polar regions and, 115–18
De Beauvoir, Simone, 104
De Coislin, Duc, 198
Dedication, 18
Degradation, 185–86
Deism, 218
Delacroix, Eugène, 107
Delay, 11–12
De Loutherberg, P. J., 106
De Maintenon, Madame, 198
Democracy, 199–209, 236
Denis, Saint, 141
Details, 12, 13
Dickens, Charles, 106, 202, 224
Distant view, 14–16, 21, 99
Dombey and Sun (Dickens), 224

Dreams, 10
  Aboriginal, 125–27
Drowsy indolence, 9–10
Dwight, Timothy, 145

E
East, 173–81, 188
Eastern (books), 179–80
Eating, taste, and culture, 46–55,
  84–85
  in China, 46, 51–55
  in Europe, 47–51
  nature as food, 229–30
Eberhard, Wolfram, 184
Eck, Diana, 112
Eco, Umberto, 139
Economic activity, 17–18
Egalitarianism, 204–205, 209
Elizabeth II, 192
Elkin, A. P., 125
Elucidation, 66–67
Emerson, Ralph Waldo, 150,
  151
Emotions, 7–13
  control of, 11–12
  music and, 93–95
  nature and, 135
  sound, aesthetics, and, 72–73
Environment and sight, 98–101
Epicritical vibrissa, 42
Eremitism, 130–31
Eroticism, 12–13, 41
Escoffier, Georges Auguste, 50
Etiquette, 198–99
Europe, 242
  cities in, 111
  food and manners in, 47–51
  medieval, *see* Middle Ages
  music in, 91–93
  odor and, 60–63, 68

Evanescent moments of beauty,
  222
Evil, 239–40
Exaggeration, 157–58

F
Familiarity, 113
Family, 237
Farms, 17–18, 110–11, 113–14,
  145
Farthest North (Nansen), 116
Fertility rites, 184
Fiedler, Leslie, 179
Fire in the hearth, 167
Firth, Raymond, 101
Flamboyance, 157–58
Flaubert, Gustave, 107
Flowers, 136–37
Flow experience, 37–38
Food, 46–55, 84–85
  nature as, 229–30
Fortescue, Sir John, 237
Foster, Susan, 39
Founding Fathers, 201, 207–208,
  209, 235
Fragrances, *see* Smell
Frailties, 238–39
France and the Sun King, 196–99,
  234, 235, 237, 239
Francis of Assisi, Saint, 136–37,
  218
Frankl, Viktor, 108
Franklin, Benjamin, 209
Franks, 191–92
Freakishness, 157
Freneau, Philip, 157
Freud, Lucian, 225
Freud, Sigmund, 55
Froissart, Jean, 105
Frontier space, 204–205

G
Galton, Francis, 168
Garden of Eden, 66, 67, 227, 229
Gardens, 44
  aromatic, 66–68
  Chinese, 67, 131–32, 215, 232
  enclosed, 67–68
  real, 67
  Versailles, 196–97, 198
Gardner, Howard, 30
Gems, 105, 139, 140
Geographic space, 174, 175
Geometric cities, 111–13
Gettysburg Address, 207, 243
Gibson, James, 42
Gigantism, 156–58
Glass office tower, 150–51
God, 66, 74, 136, 138, 192, 194,
  204, 206, 217, 218, 227, 237
  cathedrals and, 138–41
Goethe, Johann Wolfgang von,
  17
Good, 148–49, 214
Good and beautiful, 213–26
  artifacts and, 219
  artisan-artist and, 218–20
  assumptions of compatibility,
    214–18
  happiness and, 220–21
  music and, 225–26
  nature and, 217–18
  prestige and authority of art,
    220–26
  reality and, 222–25
Goodman, Paul, 10
Good Samaritan, The, 225
Gorky, Maxim, 78
Gothic architecture, 137–42
Great Day of His Wrath, The, 106
Grechko, Georgei, 15

Greeks (Greek mythology), 11,
  73, 89–90, 109, 216, 217
Green, 175, 215
Greenland, 116
Gregorian chants, 86–87
Grid pattern of towns and cities,
  148, 206
Griffin, Jasper, 11
Grudin, Robert, 22–23, 221
Guernica, 224
Guggenheim Museum, 203

H
Hale, Robert, 37
Hall of Light, 188–90
Hands, 97
Han dynasty, 127–31
Happiness, 103, 220–21
Harmony, 127–28, 186, 187, 191
  of the spheres, 89–90, 91
Hate, 10–12, 13
Hawaii, 69
Hawthorne, Nathaniel, 151
Hay, Deborah, 39
Hearing, see Sounds
Heartbeat, 72
Heartland mystique, 178–79
Heaven, 136, 137–38, 141–42
Hebrews, 74, 216
Heroes, 11
Heterogeneity to union, 205–209
Hierarchical order, 236–37
Hinduism, 66, 111
Hitler, Adolf, 239–40
Hokusai, 107
Holt, John, 25
Home, 144–45
  sublime nature and, 113–15,
    117–18
Homer, 11, 110, 113–14, 145

Hopi Indians, 185
Hopkins, Gerard Manley, 107
House of Seven Gables, The (Hawthorne), 151
Hsia-kuei, 134–35
Hsieh Ling-yun, 132
Hsun-tzu, 190
Huang Kung-wang, 132
Huang Shan, 132, 134
Hua Shan, 132
Hubris, 186
Hughes, Robert, 43
Huizinga, Johan, 105
Human beings, 240
Human cacophony, 79–81
Human sacrifice, 183–84
Hunting parks, 128–29, 131

I
Ice, 115–18
Iliad (Homer), 110
Imagination, 131–32
Incense, 64
India, 62, 65, 66, 111, 112
Indolence, 9–10
Industrial Revolution, 105–106
Infants, 21
  hearing of, 71–72
Intellectual experiences, 14
  sight and, 96–97
Intentionality, 200–201
Islam, 66, 67

J
Jackson, J. B., 148, 149, 155,
  156, 160
Jefferson, Thomas, 146, 148, 151,
  152, 157, 201–202
Jewels, 105, 139, 140
Johnston, J. R., 140

Jonas, Hans, 109
Jonathan, I., 102–103
Jones, Howard Mumford, 147–48
Judgment Day, 140–41
July Fourth, 242

K
Kalahari Desert, 99
K'ang-hsi, Emperor, 65
Kant, Immanuel, 55
Kao-yu, 133–34
Kay, Paul, 103
Kaye, M. M., 62
Keats, John, 9–10
Kilvert, Francis, 79
Kinesthesia, 35, 36–39, 154
King Lear (Shakespeare), 14
Kipling, Rudyard, 62
Kivy, Peter, 93–95
Koko, 185
Koran, 66
Kung San (Bushmen), 99
Kuo Chung-shu, 133

L
Landscapes, 114, 135
  Aboriginal, 124–27
  aerial, 155
  American, 145–46, 149, 155–
    56
  culture and, 109
  nature and, in China, 127–35,
    215
  of smell, 64–68
  of touch, 43–44
Land survey, rectangular system
  of, 148, 206
Lane, Frederic, 195
Language, 7, 74
Lawrence, D. H., 41

Legend of Good Women, The, 136
Leibniz, Gottfried, 95
L'Enfant, Pierre-Charles, 201–202, 204
Lessard, Suzannah, 108
Lévi-Strauss, Claude, 92, 99–100, 225
Lewis, C. S., 24–25, 136
Lewis, W. H., 48
Li Chi, 51
Life and culture, aesthetic in, 5–19
  aesthetic, consciousness, and emotion, 7–9
  aesthetics of the abstract, 13–16
  culture as material transformation, 16–18
  moral dilemmas, 18–19
  rage and hate, 10–12
  sexual desire, love, and eroticism, 12–13
  sleep, drowsy indolence, dream, 9–10
Light, mystique of, 138–42
Li Jih-hua, 132
Lincoln Memorial, 207
Literary art, 223–24
Literary texts, 207, 243
Livelihood, 134
London, 81
Longinus, 114
Louis XIV, 48, 196–99, 234, 235, 237, 239
Love, 12–13
Lowenthal, David, 151

M
McDonalds, 152, 160
McLuhan, Marshall, 14
Macrobius, 90
Madison, James, 208

Madness, 11
Magic, 29
Manners, refinement of, 11–12, 16, 36, 228, 239
Mark, Saint, 193, 194
Marriage of the Sea, 195, 235
Martin, John, 106
Material culture, 241
Material transformation, 16–18
Mathematics, 16
Maugham, Somerset, 56
Mbuti Pygmies, 85–86, 99, 229
Medicine, 52
Memorial Day ceremonies, 205–206, 242
Memory, 57
Mencius, 232
Metacognition, 26
Metaphors, 30, 169–71, 172
Michelangelo, 216, 220
Middle Ages, 47–48, 66, 87, 90, 105, 135–42
  the cathedral and the mystique of light, 138–42
  heavenly vault, 137–38
  nature seen closely and symbolically, 136–37
  ritual and the aesthetic-moral state in, 185–86, 191–93
Middle landscapes, 110–11, 136
Middle West, 178–79
Midnight Ride of Paul Revere, The, 154
Miller, Perry, 149
Milton, John, 67
"Mind's eye," 16
Ming dynasty, 130–33
Ming-t'ang, 188–90
Min Wenshui, 54–55
Mockery, 185–86

Molinet, Jean, 225–26
Moncrieff, R. W., 59
Monet, Claude, 107–108
Montagu, Ashley, 39–40
Montaigne, Michel de, 9
Monumental projects, 230–34
  human cost of, 230–33
  people's creativity and, 233–34
Moon, 109
Moorhouse, Geoffrey, 62
Moral and aesthetic, see Good and
  beautiful; Ritual and the
  aesthetic-moral state
Moral beauty, 240–43
Moral decisions, 219
Moral dilemmas, 18–19, 229
Moralism, 148–51, 242
Morgan, Lady Sydney, 49
Mornings on the Seine, 107–108
Morris, Desmond, 72
Mote, Frederick, 53
Mother-and-child bond, 56
Motorcars, 158–61
Mountain Bends to Man, 135
Mount Rushmore, 231
Movement, 36–39, 154–56
Mozart, Wolfgang Amadeus, 226
Multisensory experience, 165–66
Murdoch, Iris, 222–23
Music, 25–26, 72, 84–95
  as ambience, 87–88
  in China, 88–89
  as communal celebration, 85–87
  good and beautiful, 225–26
  harmony of the spheres, 89–90,
    91
  instrumental, 93–95
  listening to, 91–93
  pure music and cognition, 93–95

Western, 91–93
Mysterium tremendum, 114
Mysticism, 135

N
Nabokov, Vladimir, 24, 168
Naked Girl, 225
Nansen, Fridtjof, 115–18
National Gallery, 222–23
Natural theology, 218
Nature:
  abstract, 156
  Americans and, 203–205
  communing with, 130–32,
    146–47, 149
  culture and, 7–8, 227–29
  fear of, 60–61
  as food, 229–30
  good and beautiful, 217–18
  healthfulness of, 152
  and landscape in China, 127–35
  seen close and symbolically, 136–
    37
  silence of, 74–75
  sounds of, 71, 75–79
  state theater and, 235
  sublime, 114–15
Navaho Indians, 214, 215
Nazi Germany, 239–40
Neihardt, John, 173
Neon lights, 160
Nesterinko, Eric, 37
Newborns, 21
Newekwe, 185
New England, 180
Newness, 151–52
New York, 64, 153
New Yorker, 153
Nightmares, 10

Night Vigil, 130
Nineteenth-century novels, 223–24
Nooks, 22
Norse mythology, 116
North, 173–81, 188
North Africa, 104
Northern (books), 179–80
Notre-Dame de Paris, 136
Nouvelle cuisine, 51
Nuttall, Thomas, 75

O
"Ode on Indolence," 9–10
Odors, *see* Smell
Oglala Sioux, 173–74, 179, 180
Olfactory geography, 69
Olfactory sense, *see* Smell
On Celestial Hierarchy (Saint Denis), 141
Order:
  beauty of, 108–13
  hierarchical, 236–37
  on the land, 148
Orpheus, 73
Orwell, George, 233
Osborne, John, 13
Osgood, Charles, 169
Other-directed style, 160
Ottonians, 192
Outer space, 74, 90, 137–38
Overeating, 52
Oxford, 112–13

P
Palace of Coolness, 65–66
Panofsky, Erwin, 140, 141
Paradise gardens, 66–67
Paradise Lost (Milton), 67

Paradiso (Dante), 138
Pascal, Blaise, 73–74, 90, 137
Passing strange and wonderful, 240
Pasternak, Boris, 78
Patriot for Me, A (Osborne), 13
Pattern versus composition, 97–98
Pause, 12
Perception, 7, 166
Perfection, 186, 209
Performance, 7
Period of Disunion, 129
Permanent achievement, 186
Petrarch, 136
Petronius, 49
Photography, 147
Physical beauty, 241
Physical labor, 37–38, 45
Piazza San Marco, 193, 194
Picard, Max, 74
Picasso, Pablo, 224
Pictorial art, 224–25
Pietà, 220
Pioneering, 17–18
Plainness and simplicity, 233
Plato, 90, 172, 216, 217, 222, 225
Poets and poetry, 110, 111, 125, 127, 129, 130, 136, 186, 223
  metaphor and, 169–70
Poignancy, 30
Polar regions, 75, 76, 115–18
Politics, 182
Porteous, J. D., 62
Powys, John Cowper, 167
Practical life and aesthetic experience, 100–101
Preaching, 223–24
Pretty, 214–15

Process, 153–54
Proprioception, 35, 36–39
Proximate environment, 166
Proximate senses, 35–69
  eating, taste, and culture, 46–55
  proprioception or kinesthesia,
    35, 36–39
  smell, 55–69
  touch, 39–46
Pueblo Indians, 185
Punishment, 191
"Purifying destruction," 151
Puritans, 233
Pythagoras, 90, 225

R
Radical transformation, 229–30
Rage, 10–12
Railroads, 154–55
Ramanujan, Srinivasa, 16
Raphael, 67
Rasmussen, Steen, 81–82
Reality and art, 222–25
Recipes from Sui Garden (Yuan Mei),
  52
Reciprocity, 128
Red, 173, 175, 176
Religion, 124, 136–42, 215,
  218, 231
  see also Ritual and the aesthetic-
    moral state
Rembrandt, 224
Remembrance of things past, 23–
  25
Republic (Plato), 90
Resonance, 30
Richard II (Shakespeare), 15
Rimbaud, Arthur, 168
Ritual and the aesthetic-moral
  state, 182–209

American, 199–209
Chinese, 186–91, 215–16
France's Sun King, 196–99
medieval European, 185–86,
  191–93
mockery and degradation, 184–
  85
Renaissance Venetian, 193–96
state theater, 234–37, 242–43
violence, 183–84
Robert of Torigni, 231
Robertson, George, 106
Romanticism, 146–47, 218
Rome, 63, 80
Roosevelt, Franklin Delano, 208–
  209
Rorem, Ned, 152
Rothschild, Baron de, 49
Routine, 113
Russell, Bertrand, 16
Russia, 78–79

S
Sacks, Oliver, 58, 102–103
Sacred hoop, 173–74
Sahara Desert, 104
Sailboat in the Rain, 134–35
Saint-Denis abbey, 140
St. Louis Arch, 231
St. Petersburg, 202
San Francisco, 83
Santayana, George, 199
Santmyer, Helen, 144–45
Sargon II, 65
Savor of life, 57–58, 113
Schafer, Edward, 61
Schafer, Murray, 72, 78, 92
Schuldt, A. C., 91
Semang, 229
Sennacherib, 65

Senses, proximate, *see* Proximate
   senses
"Senses come to life," 7
Sensibility, growth in, 29–31
Sensory delight, 22–23
Sexuality, 12–13
   fertility rites, 184
   smell and, 56, 58
   touch and, 41
Shackleton, Ernest, 76
Shadows and light, 227–43
   human cost of architectural
      achievement, 230–33
   human frailties and evil, 238–40
   monumental projects, 230–34
   moral beauty, 240–43
   nature as food, 229–30
   from nature to culture, 228–29
   people's creativity, 233–34
   radical transformation, 229–30
   state theater, 234–37, 239
Shakespeare, William, 14, 15,
   218
Shang dynasty, 184
Shan shui, 128–29
Sharp, Thomas, 112–13
Shen Chou, 130, 133, 134
Shih-t'ao, 134, 135
Sight, 166
   primacy of, 96–97
   sensual and intellectual aspects
      to, 96–97
   smell and, 64
   touch and, 43, 44
   visual acuity, 99–100
   see also Visual delight and splen-
      dor
Signs, 158–61
Silence, 70–71, 73–75, 137–38
Single-minded effort, 238–39

Skin, 39–46
Sky, 108–10, 137, 138
Slonimsky, Nicolas, 95
Smell, 55–69, 72, 98
   aromatic gardens, 66–68
   bad odors, 61–62, 63
   biases and predilections, 68–69
   bonding and memory, 56–57
   city, fragrances in, 63–66
   countryside, fragrances in, 61–
      63
   educated nose, 58–59
   extraordinary sense of, 58
   loss of sense of, 57–58
   nature, fragrances in, 59–61
   periodic, 62–63
   savor of life, 57–58
   sexuality and, 56, 58
   sight and, 64
Smyth, R. Brough, 122
Social harmony, 148
Solipsism, 222–23
Solomon, 65
Songlines, 126, 127, 177
Sontag, Susan, 12–13
Sounds, 70–95
   absence of, 70–71, 73–75
   biological rhythms, 71–72
   cities, 81–84
   "colored hearing," 168
   emotion, aesthetics, and, 72–73
   everyday, 71
   human, 79–81
   infants and, 71–72
   music, *see* Music
   nature and, 71, 74–79
   power of, 95
   voices of, 71, 73
South, 173–81, 188
Southern (books), 179–80

Southworth, Michael, 82–83
Space, 74, 90, 137–38
  frontier, 204–205
  geographic, 174, 175
  symbolic, *see* Symbolic space
  vagueness of, 200
Spain, 63
Speech, 7
Stained glass, 140, 141, 224
Stark, Freya, 221
Stars, 109, 110, 138
Stasis, 238
State theater, 234–37, 239, 242–43
Statue of Liberty, 231, 242
Stein, Gertrude, 153
Stevens, Mark, 225
Stilgoe, John, 145
Strange and marvelous other, 23
Strip, the, 158–61
Styron, William, 226
Sublime nature:
  home and, 113–15
  landscape as, 128
  polar regions and, 115–18
Suburbs, 131
Suger, 140, 141
Sui dynasty, 188
Sullivan, Louis, 203
Sullivan, Michael, 133
Summa Theologiae, 139
Sun, 109, 138, 196
Sung dynasty, 130, 133
Sun King, 196–99, 234, 235, 237, 239
Sunsets and sunrises, 108–109
Suso, Heinrich, 137
Su Tung-p'o, 130
Symbolic space, 172–81
  American, 178–81
  Australian, 177–78
  Chinese, 174–77, 179, 180, 187–88
  Oglala Sioux and, 173–74, 179
Symbols, 171–72
Synesthesia, 167–68, 169
  children and, 170
Synesthetic tendency, 168–69

T

Tactile aesthetics, 39–46
T'ai-tsung, Emperor, 191
T'ang dynasty, 186–91, 201
Taoism, 67, 111, 128–30, 135, 176, 215, 216, 230
Tao Yuan-ming, 111, 129
Taste, 46–55
Ten Thousand Ugly Ink Dots, 134
Theatrical model of human reality, 7
Thomas of Celano, 137
Thoreau, Henry David, 150, 151, 152
Thubron, Colin, 46
Tikopia, 101–102
Time and space, 91–92
Timelessness, 22–23
Tolstoy, Lev, 38, 62, 78–79, 216–17, 223
Top-down process, 206
Topographic features, 13
Totalitarianism, 220
Touch, 39–46
  eating as, 46
  landscapes of, 43–44
  reality, beauty, and, 45–46
Tourists, 157–60
Towns, 144–45, 148
Traveling Upriver in Midwinter, 133
Ts-ui, 53–54

Tung Chi-ch'ang, 132
Tung Chung-shu, 176
T'ung Yuan, 133
Turing, Alan, 16
Turnbull, Colin, 85–86
Turner, Frederick Jackson, 205
Turner, J. M. W., 222
Twilight, 101–102
  gemlike fire versus, 104–106
Twilight in the Wilderness, 106

U
Uganda, 77
Ugly, 215
Union of diverse elements, 205–209
Updike, John, 12, 71

V
Vagueness, 200, 209
Venice, 88, 193–96, 235
Venus, 99–100
Vergerio, Pier Paolo, 196
Versailles, 196–99, 201, 232, 234, 235, 238, 239
Violence, 11
  ritual, 183–84
Virgil, 110
Visual art, 27–29
Visual delight and splendor, 96–118
  beauty of order, 108–13
  color, 102–104
  composition versus pattern, 97–98
  environment and culture, 98–102
  gemlike fire versus twilight, 104–106
  ice, 115–18

intense and joyful seeing, 106–108
Voices, 71, 73
Von Simson, Otto, 141

W
Walkabouts, 125
Warner, Lloyd, 205–206
Washington, D.C., 201–203, 207
Washington, George, 201, 204, 207–208
Wasserman, Robert, 102–103
Watson, G. N., 16
Webster, Noah, 151
Wechsler, Lawrence, 95
Weil, Simone, 12, 45
Wen Chen-ming, 131
West, 173–81, 188, 200
Western (books), 179–80
White, 173, 175, 176
White, Lynn, 231
Whole, 216
Wilderness, 60–61, 74–75, 77, 127, 130
  in America, 145–47, 149, 218
  good and beautiful, 217–18
  Romanticism and, 146–47
William the Conqueror, 192
Wilson, Edmund, 154
Winter Evening, 135
Wittgenstein, Ludwig, 92–93, 220, 226
Wonder, capacity for, 23
Wood, Grant, 154
Words, 207, 243
Wordsworth, William, 223
Wright, Frank Lloyd, 203
Wright, Joseph, 106
Wu, Emperor, 129
Wu, Empress, 190

Y
Yellow, 173, 175
Yellowstone, 157
Yin and yang, 44, 52, 132, 174,
    176, 215
Young, James, 80–81
Yuan dynasty, 131, 135
Yuan Mei, 52

Z
Zaire, 77, 85–86
Zuckerkandl, Victor, 86
Zuñi Indians, 185, 214–15

A B O U T   T H E
A U T H O R

Yi-Fu Tuan is the J. K. Wright and Vilas professor of geography at the University of Wisconsin-Madison. His books include *The Hydrological Cycle and the Wisdom of God* (Toronto), *Topophilia* (Prentice-Hall), *Space and Place* (Minnesota), *Landscapes of Fear* (Pantheon), *Segmented Worlds and Self* (Minnesota), *Dominance and Affection* (Yale), *The Good Life* (Wisconsin), and *Morality and Imagination* (Wisconsin). He is currently working on cosmopolitanism versus ethnicity.